Nature's Trajectory

Also by Louis Perry:

Jefferson's Scissors:
Solving the Conflicts of Religion
with Science and Democracy

Atheist, Friend or Foe?
What Christians Should Know

Thank Evolution for God:
The Role of Nature
and God in Evolution

Nature's Trajectory

The Creation of the Heavens, Earth, Man, and His Gods

Revised

Louis Perry

Copyright © 2018 Louis Perry
All rights reserved.
ISBN: 9781077444164

Dedication

This book has been made possible by the help of many. Special thanks go to Dr. Tom Bond, (retired) provost, Revelle College, University of California San Diego, for offering the course on Conflicts of Religion with Science and Democracy; Dr. Scott Hestevold, chairman, Philosophy (retired); and Dr. Norvin Richards (retired), University of Alabama, for their philosophical insights into the evolution of morality with and without gods and to Rev. Laurel Grey, Lutheran minister (retired), for his insights into progressive Christianity. And thanks to Joan Fisher (my partner) for her unwavering support.

Appreciation is also given to the scientific community for their research exploring Nature on which this book is based. Further, for their continuing efforts in the face of ongoing religious and political impedances to their research discovering Nature's laws and how they really work.

Content

I. Introduction 1
II. Authorities 7
   Nature 9
   States 10
   Gods 11
   Alternatives 14
   Law Hierarchy 16
III. Scientific Discoveries 17
   Technology 23
   Astronomy 26
   Physics 33
IV. The Universe 41
V. Biological Life 47
   Early Life 53
   Vertebrates 55
VI. Mental Constructs 75
   Gods Appear 78
VII. Conflicts 87
   Nature and God 89
   State and God 98
   Nature and State 106
VIII. Summary 115
Appendix A 121
   Summary of Four Forces of Nature 121
Appendix B 123
   Notes on the Ten Commandments 123
Appendix C 125
   Morality as Viewed by Heisenberg 125
Index 127

Figures

1. Galileo's Observations of Heavenly Bodies
2. Growth of Optical Telescopes
3. Cosmic Microwave Background Radiation
4. Black Hole Image
5. Evolution of the Early Universe
6. Evolution of Earth and Life
7. Evolution of Life on Earth
8. Evolution of Vertebrate Life
9. Coelacanth—Living Fossil
10. Chimpanzee Family
11. Evolution of Hominins
12. *Australopithecus* Fossil Footprints—The Couple
13. *Australopithecus* Fossil—Selam
14. Evolution of Awareness of Spirits and Gods
15. Epic of Gilgamesh—Tablet, 2100 BCE
16. Creation of the Universe—God's Timeline
17. Creation of the Universe—Nature's Timeline
18. Nature's Trajectory in a Nutshell

Foreword

The background story is that we are on a small planet orbiting an ordinary star hurtling through a vast expanding cosmos, a relic from Nature's big bang 13.72 billion years ago. How did we get here and where are we going? To answer these questions, we will explore what Nature is telling us about the trajectory of the expanding universe from the big bang and the creation of the Earth, man, and his gods and address *what* has been created and *how* it happened.

Many people will ask, "What about my god being the creator?" This question is answered in the course of examining man's evolution and his subsequent invention of narratives about gods. We will address *when* gods appeared and *what* do the many religious narratives tell us about god's creation. We will also address why the narratives present conflicts with has been discovered about Nature.

We have learned about Nature through scientific discoveries in physics, chemistry, and biology that have given us insights into Nature's creation of the universe, Earth, and biological life. From these studies, we have learned about the evolution of *Homo sapiens,* namely us, and our capability to conceive of and write narratives that describe supernatural creations by the gods. And indeed, *Homo sapiens* have conceived of many gods, each with a creation story. But for our discussions we will use only the Christian God narrative as an example. Thus, we have two creation stories to discuss, Nature's and God's. As expected, when comparing these creation stories based on different authorities, there are many conflicts.

Nature's information about the evolution of the universe and man continues to increase, so the conflicts with religious narratives will continue to deepen with time if the roles of science and religion in society are not clearly understood. In our democracy we will continue to have both science without gods and multiple religions with their gods for a long time, so it is important to define compatible roles for each in order that we jointly attack the global problems facing us. Having science, the state, and religion work together is a matter of our survival.

I. Introduction

What is our genesis? Are we special in this universe? What roles in our creation are played by Nature[1] and by man's gods? Is there not a natural human inclination to ask questions about the evolution of our vast universe, the Earth, and its diverse biological life, including man? These questions are addressed by examining Nature's trajectory from the big bang to today.

To have confidence in any answers to these questions, we must identify the authority of the information used in answering them: is it based on Nature's laws, which are supported by experiments and scientific data, or is it based on interpretations by clergymen of God's laws taken from ancient "holy" narratives written by believers centuries ago who had very little knowledge of science?

We receive information about the world from many sources. As children, we become aware of the existence of Nature through hands-on experiences with objects: sounds, lights, words, books, and interactions with our family, friends, and others. As we grow older, we may go to church and learn about a god and his laws and our obligations as a believer.

We also go to school and begin to learn about Nature's laws, such as Newton's law of motion and Nature's language, mathematics, from our teachers. Later, in many of our jobs, it is necessary to learn more about Nature's laws to make things work or to build things, like a bridge, where some information will be critical—what is the strength of steel beams and other material, the rate of oxidation, the thermal expansion, etc. In the medical world, to heal people, we need information on drugs, and so on.

---

1. "Nature" is capitalized as its laws are universal and unique. "God" is capitalized when referring directly to the Judeo-Christian God.

For this information, we turn to natural science with its mathematical theories to understand how things work in our everyday life because science gives answers that work predictively, although not always perfectly for they are administrated by man. We can depend on natural science to the same degree regardless of the religion of the person we ask—whether they are Hindu, Christian, or Atheist—for information about Nature comes from the same universal source so that everyone works from the same body of scientific knowledge.

In universities, we learn[2] things about a broad range of topics, including the many religions of the world and their similarities and differences. We learn about more abstract theories of Nature, such as the movement of stars (Newton's law) in astronomy classes and the interrelationship of energy and mass (Einstein's $E = mc^2$ equation) in physics classes. And in biology classes, we learn about biological evolution (Darwin's theory of natural selection). Examination of fossilized bones gives an indication of what animal life looked like in the past. The teeth reveal its diet and the skeleton its shape. Tools, art, and other artifacts found at sites tell us about its culture. Further, DNA, which can be extracted from some of the fossil bones, gives us comparative information about the genetics of the species.

Most American children are told about gods by family and through participation in local churches and other houses of worship that relay religious narratives and dogmas to follow. Christians learn that their religious narrative describes a supernatural[3] God who created the universe, the planet Earth, and all life, the first being a man and a woman called Adam and Eve. However, the Judeo–Christian narrative does not describe the evolution of our ancestors before *Homo sapiens* (chimpanzees and the many other species from earlier times) or a universe with stars and galaxies dating back billions of years; instead it contains supernatural stories of instant creation of humans, worldwide floods, and miracles with dates of thousands of years ago that are in conflict with what is learned in science classes.

The understanding of Nature by scientists has come from the many scientific discoveries accomplished by a persistently curious and sometimes lucky scientific community. The discovery of some[4] of Nature's

---

2. Religion-based universities may filter out some science information that conflicts with their dogma.
3. The word "supernatural" is used to describe information that is outside the natural world; that is, information not based on Nature's laws and unsupported by evidence.
4. There may be more to be discovered.

laws has given us an initial insight into our genesis and a basis for explaining the continuing unfolding of the universe. The data that have been gathered are far from complete, but enough has been learned to give a compelling glimpse of Nature's trajectory from the big bang to us, *Homo sapiens*. The view of our creation by most scientists and this author is that Nature's laws created the big bang and it was the genesis of our universe.

The picture currently emerging from research is that of a universe created some 13.72 billion years ago from an event, the big bang. It is the singular event and the common ancestor of all forces, particles, matter, and things in our expanding universe with its myriad cosmic bodies as well as our Earth with its biological life, including man (*Homo sapiens*). With man we also have the stories he has conceived about supernatural gods.

With religions we have different stories about creation than that of Nature. The differences between Nature's creation of the universe and the Judeo-Christian God's creation as envisioned by early believers and described in Jewish and Christian biblical narratives are major. To begin with, the Judeo-Christian narrative[5] dates the genesis of the universe to about 6,000 years ago, a far cry from the 13.72 billion years estimated from studying Nature. Belief in this religious creation story requires rejection of some of the scientific data that have been discovered. These data are from the big bang's creation of an expanding universe, to formation of Earth, to evolution of life on Earth, to the evolution of *Homo sapiens*, and to the evolution of his mental capabilities to the point where he can conceive of supernatural gods. These conflicting accounts present an unbridgeable foundational gap between the two creation stories. In the vernacular of the space program:

Houston, we have a problem.

This is not a new problem; it emerged at the beginning of the scientific revolution when Nicolaus Copernicus in 1543 published his heliocentric theory challenging the Christian dogma of a geocentric view of Earth's position in the heavens. The Vatican geocentric theory was dogma without proof. But that was to be challenged when observations of Jupiter's moons were first made by Galileo Galilei in Rome and led him to embrace the Copernican heliocentric view of Earth and the heavens.

---

5. Although Christianity is used as the reference religion, there is a spectrum of acceptance of science among the different denominations: from fundamentalists, who believe in biblical accuracy, to progressives, who accept most of natural science theories.

This violated the Vatican's dogma of the universe, and in 1633 the Vatican accused Galileo of holding heretical views based on his description of his observations of the heavens. Galileo's famous reply to the Vatican's charge was:

> *The Bible tells us how to go to heaven, not how the heavens go.*

The Vatican would not accept Galileo's *how the heavens go,* an argument that separated science from religion, and instead kept to its policy of having their God tell how the universe and everything goes. The Vatican had the power of the church and state over the information, but they made a most unfortunate decision. If they had accepted Galileo's attempt to separate the two views, there would be far less conflict. But even after years of research that supports Galileo's scientific view, some churches still insist that the Bible tells believers not only *how to go to heaven* but also *how the heavens go.*

Our discussions will focus on *how the heavens go*; that is, the science of the cosmos. What churches tells us about *how to go to heaven* is not a matter of science and will be left to church stories.

If Nature and not the God of the Bible is accepted as the causal agent for the creation of the universe, then we accept scientific observations and experiments as the authority on which to base our understanding of the genesis of the universe and the creation of biological life on Earth. This life is widely diversified and includes the species *Homo sapiens*—us—a species that has evolved an inquisitive mind with the capability not only to seek an understanding of the mysterious and improbable things we see on Earth and in the heavens but also to devise concepts of supernatural gods. *Homo sapiens* have very fertile minds.

How the heavens go (Nature's laws of the universe) has been addressed through scientific observations and experimentations on Nature. With this information scientists have constructed theories of the trajectory of Nature's universe from the time of the big bang billions of years ago to today. These theories allow a description of the evolution of *all things,* including stars, galaxies, the Earth, and man and his god concepts. But as new discoveries are made the theories are expected to change in the future.

Consideration of *all things* would generate a far too large number of topics to cover in this small book, so the links from Nature's creation as the causal agent of man and his gods are summarized. There are many science books available (many are noted in this book) for those who wish to dig deeper and learn more about the breadth and depth of the supporting science that has been discovered.

To describe the evolution of our understanding of Nature, we need to go back in time and outline the emergence of scientific information describing the universe and biological life and the construction of supernatural-god narratives over the 13.72 billion years since the big bang. This is done in four *once upon a time* stories that outline sequences of scientific discoveries from which scientists have gathered information about the universe, biological life on Earth, and gods conceived by man:

- ***Once upon a time***, there was no time, no space, and no universe. Nature created a big bang that produced space–time and the universe with many galaxies, stars, and planets, one of which is Earth.
- ***Once upon a time***, on Earth, Nature created biological life from the chemicals (cosmic dust) on Earth that evolved into complicated life forms, one being the species *Homo sapiens*.
- ***Once upon a time***, *Homo sapiens* evolved the mental capability to explore Nature and to conceive of complex stories, some with supernatural god characters.
- ***Once upon a time***, from the many gods conceived by tribes the Jewish tribe selected their God and described him in a religious narrative, the Torah. Subsequently a Christian narrative, the Bible, evolved and described the Christians' God.

The first two stories were triggered by improbable events: the first, the big bang creation of the universe and Earth, and the second, the creation of biological life on Earth from cosmic dust with a self-replicating molecule that subsequently evolved into the species *Homo sapiens*. The third story highlights the growth of the *Homo sapiens* brain and its creative ability to tell tales with supernatural gods. The fourth story describes *Homo sapiens'* invention of a Jewish narrative, from which a few centuries later a Christian narrative evolved with the Christian God. The first two were low-probability events of Nature. The third and fourth were not improbable, but events created by *Homo sapiens* in the course of their evolution.

So here we are, fortunate historians poring over tons of data from observations of Nature and from religious narratives trying to write the story of how man and his gods all came about. Many steps were taken in this evolutionary journey. Key scientific advances that give us a foundational view of this evolution are discussed. Religious views on the

Jewish and Christian Gods are taken from the Jewish and Christian narratives.

There is a wide spectrum of religious beliefs about gods and what believers' narratives say about gods. These differences are not debated here, for it is up to each person, if they choose, to select the supernatural religion and god from the denomination of their choice and to debate with other believers which god is *the only god*. All religions and their believers are free to hold whatever views they choose about creation of the universe and man. This book's religious discussions are limited to the Christian religion and its narrative to provide examples of their God's words and laws. Also, for brevity, the evolution of Western science is the focus; not included are the considerable contributions made by Islamic and Eastern societies in science, mathematics, and religions.

The *once upon a time* stories and supporting material are discussed in the following chapters:

- Chapter II - Authorities – A summary of the informational authorities of Nature, states, and gods as well as sources of misinformation, such as alternative facts.
- Chapter III - Scientific Discoveries – An outline of key scientific discoveries and the scientists who made them in the disciplines of astronomy, physics, and evolutionary biology that have propelled our scientific knowledge.
- Chapter IV - The Universe – A summary of the present scientific understanding of the big bang's creation of the universe and Earth.
- Chapter V - Biological Life – A summary is outlined of the creation of biological life on Earth, from the first self-replicating molecule to the evolution of vertebrates and modern *Homo sapiens*.
- Chapter VI - Mental Constructs – A summary of the evolution of the *Homo sapiens* brain's capabilities to produce complex social concepts, including gods (religions) and governments (states).
- Chapter VII - Conflicts – A discussion of the conflicts that have arisen between Nature's laws and the laws of man's social constructs—states and religions (gods).
- Chapter VIII - Summary – A summary of the scientific information supporting Nature's creation of the universe, Earth, man, and his gods. An overview of some of the possible future trajectories Nature may take and their impact on mankind is given.

II. Authorities

Before having a discussion, it is important to know the authorities for the information being used, for all authorities may not have the same rules and bases for the information. Most information we never hear about. The information we hear is either accepted or rejected, depending on a person's unique set of preferences and values. For example, far different conclusions will come from discussing the creation of the universe if one uses information from any of the many supernatural religious narratives instead of that from scientific texts.

There are several formal authorities for the information we use daily: Nature (science), the government (state), and god (religion) if we are believers. There is a fundamental difference between the universal scientific authority based on observations of Nature and the authorities of states or religions, which are based on social constructs by man and are not universal. There are many other informal social authorities that one may use, but for discussion we will focus on the three, science, state and religion, that have evolved and recorded laws.

Nature's laws of physics discovered by scientists are universal and underpin our physical activities on Earth as well as those of faraway galaxies. Throw a ball, shoot a gun, or launch a rocket, and the trajectory of the projectile is described by Newton's law of motion; walk in the woods on a camping trip guided by a magnetic compass's pointer, which aligns with Earth's magnetic field as described by Maxwell's equations; or read the time on your watch dial illuminated by the light of a photoluminescent screen activated by particles from the decay of a radioactive element by the weak force as described in nuclear physics. Further, we cook our hot dogs over fires that give us heat and light as explained by Nature's laws of chemistry and physics. These and the other laws of Nature are universal and used every day, everywhere, by everyone doing anything.

Of the three authorities, science has a unique position in our world, for it uses Nature's laws to describe man's evolution, from the chemical elements on the Earth to the self-replicating DNA molecule in each of our cells. Further, it describes our little planet as well as the largest galaxy billions of stars light years distant for they all function according to the same laws of Nature. The language used to describe Nature's laws is mathematics. Our understanding of Nature and its laws is based on the mathematical equations in scientific theories.

The authorities of the state and religions (gods) do not use mathematics as their language; instead, they use the local language with which they have constructed many narratives and social theories, including the laws of their state and religions. Without reference to the scientific laws describing Nature, religious and state narratives and their laws have no constraints and can describe supernatural concepts as well as natural ones. This leads to many possible conflicts.

The study of Nature by scientists has revealed that the evolution of all species of biological life on Earth is described by Darwin's theory. Our species, *Homo sapiens*, has evolved not only physically but also mentally with the capability to conceive of stories with supernatural concepts. Two major social concepts have been invented by man: governments (states) and religions (gods). Governments vary from simple powerful rulers to kingships to sophisticated democracies. Religions vary in that they have different supernatural gods, angels, devils, etc. Each of these two social constructs has a narrative that includes laws to be followed and an organization to manage believers.

In our daily lives as citizens of our country, we interact with not only Nature's laws—for they are unavoidable—but also state laws and, if we are believers, religious laws ascribed to a god. Further, we may also decide to follow the rules and taboos of informal social groups, those of our family and interest groups. State laws are constructed by the people in democracies, and in nondemocratic governments they are dictated by the leaders and enforced with or without consent of the people. God's laws are conceived by believers, and promulgated by godly organizations (churches), recorded in religious narratives (Bibles[6]), and are valid only to people who choose to be believers. God's laws differ from religion to religion and are selected and enforced by the authority of each church's organization that interprets god's words and declares which are "holy" words of a religious narrative. For simplicity, our focus is on the interactions of these three major authorities affecting our lives—Nature (science), the state (governance), and God (religion). Some laws of these different authorities conflict and present problems (see Chapter VII).

Compounding information selection problems is the misinformation presented by "alternative authorities." Our brains are besieged daily with information from questionable or unknown authorities projected by con

---

6. Richard Elliott Friedman, *Who Wrote the Bible?* Simon & Schuster, 1987.

men, liars, ignorant people, and others, so we must continually check the source of the information. Each of us has the job of accepting or rejecting each piece of information on which we base our discussions and decisions: is it based on Nature's laws, state laws, or god's laws? Care should be taken to reject "alternative information," for it is highly probable that it is an ad hoc invention or a lie by a perpetrator trying to sell you something with no recognized supporting authority.

The authorities for Nature, states, and gods are outlined below. The lack of authority for alternative information is also discussed.

Nature

Nature's laws extend over all matter and forces in the universe, from the elementary subatomic quarks and neutrinos to compound particles (neutrons, protons) and to the largest galaxies in the universe. Matter is acted on by four forces: the strong force, the weak force, electromagnetism, and gravitation. The most familiar forces are gravitation and electromagnetism. Gravity explains the fall of an apple on Earth and the Earth's orbit around the Sun, and it provides the force predicted by Einstein for constructing black holes. Electromagnetism is the force carried by the electric and magnetic fields; electric fields provide the power for electric motors in cars and magnetism of the Earth is seen in simple magnetic compasses. The strong and weak forces are out of our sight for they act at the very small nuclear particle level; the strong force holds ordinary matter together and the weak force is responsible for nuclear particle decay.

There are two other forces we know little about: dark matter and dark energy. Dark matter interacts primarily with gravity and we observe its interactions holding galaxies together, but not much else is known about it. And there is dark energy, a force also not well understood that permeates all of space and has, over the last four billion years, caused the acceleration of the expansion of the universe that has been observed.

The discovery of some of Nature's laws by scientists has placed the scientific community as the authority on the laws. Our knowledge of Nature is limited, but new discoveries continue to expand our knowledge.

The authority of Nature's laws has been found to be based on the following:

- Source – Nature; its laws are described by mathematical theories.
- Universal – They apply everywhere in the universe.

- Absolute – Nothing in the universe affects them.
- Omnipotent – Everything in the universe must comply with them.
- Unique – There is only one set of Nature's laws.

Nature's laws are based on verified scientific information that is open to change in the future as scientists dig deeper into understanding Nature. In their task to understand Nature, scientists have made mistakes and will continue to make them, but the scientific method, which mandates verifiability of the evidence, allows for the correction of mistakes when the data so indicate.

When scientists make mistakes formulating or using Nature's laws there are no human punishments, only observations of errors to be corrected in order for things to work. Participants are rewarded in their search for Nature's laws by being able to accomplish tasks that work and others that could not do before and by the personal satisfaction of contributing to advancing the international scientific community's knowledge of Nature. In some cases, the scientists win prizes, such as the Nobel Prize for physics, chemistry, etc. Participants not complying with Nature's laws fail in activities, such as building bridges that collapse when the structure is too weak to carry the load or designing rockets that explode due to design mistakes.

States

It is assumed that the tribal leader of early *Homo sapiens* was the alpha male, as we have observed with our ancestors, chimpanzees[7] and gorillas. However, as the number of humans increased it became necessary for tribal leaders to manage larger groups, so organizational hierarchies, operational laws, dogmas, and protocols evolved. These were recorded after writing became available; an example is King Hammurabi's Code. Religions had their own laws and commandments. States were usually theocracies that combined the mysterious power associated with god's religious laws with state secular laws.

Over time, in addition to theocracies, many different types of states have been conceived and tried: monarchies, democracies, republics, dictatorships, communist states, socialist states, fascist states, and so on. The rights of individual citizens began to appear as citizens confronted their kings and churches. In democracies, laws are secular and apply to

---

7. But not bonobos, which are led by females.

citizens of all religions. Religious believers can also choose to accept additional laws of their religion.

The American Revolution produced our democracy, whose authority is based on a secular written constitution constructed by its citizens. Its provisions include the separation of religions and their laws from the state's laws. This is critical in a democracy that has citizens of many different religions, for when governments authorize any one religion it gives power to that religion's laws and imposes these laws on all citizens, even those who are nonbelievers or believers in other religions.

The laws written in our Constitution are from the people, by the people, and for the people. These laws are not static, and change occurs when the collective social cultures of the citizens change and citizens vote, through their representatives, to make changes in the laws. The Constitution is constructed so that it can be modified by amendments at the will of the people, for the founding fathers knew that they could not predict the future at the time the Constitution was written and that changes would be required.

Over time, major changes in our constitutional laws through amendments have occurred. Some examples: laws supporting slavery were eliminated, women were given the right to vote, and all persons were granted the right to marry the person of their choice. The cultural changes from such laws took a long time, but the state laws were changed.

In summary, we can say the following about the authority of our state laws:

- Source — They are social laws constructed by man.
- Universal – No; they apply only to citizens of one state.
- Absolute – No; they can be changed by citizens.
- Omnipotent – No; they are not valid in other countries.
- Unique – No; each state has its own set of laws but some laws are used by other states.

Gods

Supernatural spirits and gods were conceived by man to answer questions: "Why do mysterious things (lightning, volcanic eruptions, floods, deaths) happen?" "Why am I alone in the world?" "What happens to me after death?" These things were initially explained by reference to supernatural spirits that had been invented. Early shamans and writers of religious narratives had vivid imaginations and without access to science information that would allow them to discriminate between natural and supernatural forces or events, they invented supernatural spirit and god concepts and used them in their religious narratives.

Such narratives listed dogmas and laws to direct the worship of their gods, including what should be worn, what should be eaten, etc. The narratives also included secular laws used by the clergy to direct the congregations. The laws were declared by the church leaders to be from god and, therefore, sacred and unchangeable, for how can god be wrong? However, cultures change, and new discoveries create conflicts with godly pronouncements and commandments for religions.

Every religious group has its religious narrative, each with a different god and a different set of laws with differing punishments for disobeying. Religions evolved and many religious concepts and laws were adapted and used by later religions. The Code of Hammurabi established the presumption of innocence. The Zoroastrianism of early Persians had one god, Mazda, which has influenced subsequent religions with their concepts of heaven and hell, free will, the fight between good and evil, and judgment after death. The Torah, the holy book of an early religion, Judaism, borrowed from these earlier religions; an example is the creation story of the heavens and the first humans with their Adam and Eve story. Another example of borrowed stories is the flood myth, which came from an early Sumerian poem, *The Epic of Gilgamesh*. The Jews retold the flood story in the Torah, describing the Jewish God being upset with the evil in the world and sending a worldwide flood to kill everyone except Noah and his family. Later the Christians incorporated the same flood myth into their holy book, the Bible.

The Christian narrative gives God's laws for religious guidance and outlines punishments for those who disobey, which range from death to expulsion from the religious community to punishments in the supernatural world, such as exclusion from heaven or a sentence to eternal life in hell. With no basic source to say what is Christian and what is not, different Christian denominations have different interpretations of the punishments offered in biblical texts.

Some biblical laws are secular social constructs based on observations of man's common morality that aided their tribal survival. These laws provided useful guidance for other societies before the Torah and the Bible were written. Some of these laws have passed the test of time and continue to be acceptable today. They are used by religious and nonreligious societies. However, some of the social laws (and commandments) in religious narratives were based on social cultures of a time past (2,000 to 3,000 years ago) and are now unacceptable because of changes in social culture and morals in societies.

Some examples of social and religious laws in the Christian Ten Commandments that are acceptable and unacceptable today are:

Social laws acceptable today:
- Prohibition of stealing (Exodus 20)
- Prohibition of murder (Exodus 20)

Social laws unacceptable today:
- Death to unchaste women at time of marriage (Deuteronomy 22:20–21)
- Sanctioning of slavery (Leviticus 25:44)

Religious laws acceptable today for Christians:
- Worship only God (Exodus 20)
- Make no graven images (Exodus 20)

Religious laws constitutionally unacceptable today:
- Death for blasphemy (Leviticus 24:16)
- Death for nonobservance of the Sabbath (Exodus 31:14)
- Death to witches (Exodus 22:18)

The out-of-date, unacceptable biblical laws[8] pose dilemmas for Christians. Since they are taken from the biblical text and as such declared to be God's holy words, they must be correct and everlasting, but they obviously are not.

The Ten Commandments (see Appendix B) is an example of a mix of religious and socially acceptable laws that include four religion-specific commandments (laws) and six social commandments (laws). The six general social commandments, such as "Thou shall not steal," are humanistic secular morals that are in general use by other societies and religions. The four commandments that are religion-specific, such as "Do not take the Lord's name in vain," are not useful outside the specific Christian community.

---

8. Blasphemy as a law is still around in some countries today. For example, Pakistan currently sentences people to death for blasphemy.

The overall Christian moral authority is based on the biblical text that was written and assembled by believers 2,000 years ago. Since then believers have commented on the biblical narrative in light of changes in societies, but there is not a Christian authority to update the biblical narrative or revise their laws, even those that have become culturally out of date. It is left to individual Christian denominations to observe, enforce, or neglect such religious laws.

In summary, the following can be said about the authority of the Christian God's laws:

- Source — They are social constructs described in religious narratives written by man.
- Universal — No; religious laws apply only to believers in that religion. Secular humanistic laws included in them are used by some, but not all, religions and societies.
- Absolute — No; they are interpreted differently by believers.
- Omnipotent — No; they do not apply to all peoples, only believers. However, religious narratives provide guidance and emotional support to many, empathy for those in need, and hope for those who seek words of support.
- Unique — No; other religions and states have many comparable religious and secular humanistic laws in their narratives.

Religious believers declare that there are events and concepts that cause them to feel awe, as if they have witnessed something otherworldly, amazing, and possibly frightening, but satisfying. People with deep religious feelings have expressed that interacting with the characters and stories in the Christian narrative gives them this feeling, as if they are in the presence of a supernatural force that they identify with their God. Some have used such personal feelings as proof of their God's existence without understanding that personal feelings are unique to the individual and not repeatable and therefore do not constitute proof.

Alternatives

Alternative "facts" that are unsupported by recognized authorities are often put forth by individuals and organizations. In effect, such alternative facts are stand-alone pieces of misinformation fabricated by people for political, economic, or self-serving reasons. The following can be said about alternative "facts":

- Source – Created by man's inventiveness to create texts without support from the laws of any authority.
- Universal – No; they apply only to the speaker and accepting recipients.
- Absolute – No; they are interpreted differently by recipients.
- Omnipotent – No; they do not apply to all peoples, only to accepting recipients.
- Unique – No; they reflect misinformation or lies often repeated.

At times, people have been guilty of using alternative facts in arguments between science, state, and religious laws. An example comes from the North Carolina legislature, which, after receiving a scientific report on global warming about a potential sea level rise on their coast, barred state and local agencies from developing regulations or planning documents that accept science's anticipated rise in sea level. The North Carolina law is based on a false political alternative "fact" that the sea level will never rise. Upon hearing about the new law, a comedian commented:

If your science gives you a result that you don't like, pass a law saying the result is illegal.

A recent example of the use of alternative facts is a statement by President Trumps press secretary, Kayleigh McEnany. Trump is a science denier, so, following his lead, McEnany denied the medical facts about an increase in the spreading of the coronavirus in relation to whether children should return to a normal classroom environment and announced:

We should not let science stand in the way of our children's education.

These examples are unfortunate, but when leaders in the country deny science and laws like this are enacted on a large scale across the country, we could be entering a *post-truth* era driven by political goals. Lee McIntyre[9] noted:

Post-truths amount to a form of ideological supremacy, whereby its practitioners are trying to compel someone to

---

9. Lee McIntyre, *Post-Truth,* Cambridge, MA, MIT Press, 2018.

believe something whether there is good evidence for it or not.

. . . In a post-truth era "alternative facts" replace actual facts and feelings have more weight than evidence.

Clearly, in a *post-truth* era using alternative facts, Nature's laws, constitutional laws, and even religious laws are at risk. This is an era that we should surely work to never be in again. When faced with alternative "facts," each of us has the burden of recognizing and rejecting these fabricated fictions and lies that have no supporting authority. Politicians do not have the authority to violate the laws of Nature or the state (the Constitution) and use alternative "facts."

Law Hierarchy

In our search for authorities to support the information we are using for decisions, we must be aware that there is a hierarchy of authorities. At the top are Nature's laws, which prevail over the state's laws and religious (God's) laws. Secondly, state laws also prevail over religious laws. Not accepting this hierarchy of superiority leads to conflicts. Thirdly, religious laws are for specific religious organizations. Alternative "facts" should be rejected for they can only lead to conflict.

However, much of the information we encounter daily is the reverse of this hierarchy. We are usually awash in rumors and alternative information. If we are religious, there are many daily reminders of religious protocols and God's laws. We know less about state laws applicable in a situation and some research may be needed. To find Nature's laws requires even more effort to apply. Thus, it usually takes some work to apply information from the right authority on decisions we are making. For critical decision it is often necessary to do the hard work. This is particularly true of applying Nature's laws that challenge state and religious laws that have been around for a much longer time. Chapters II to VI describe the emergence of some of Nature's laws from scientific discoveries made over the last few hundred years. Chapter VII discusses some example conflicts that have arisen and, in many cases, continue between science and the state and between science and religion.

III. Scientific Discoveries

Scientific discoveries are a natural outgrowth of man's curiosity about the world and what makes things work around him. Early thinkers and philosophers from many countries have studied the fundamental questions about our existence and values. It has taken time to develop the source material and the experimental tools for supporting their search for answers to these questions, but scientific discoveries have given scientist new tools needed for research.

After writing was invented, man started to record his observations of Nature and systematic science began. Writing began when the Sumerians developed the earliest writing system (cuneiform), and around 5,000 years ago and with this primitive script they recorded movements of celestial bodies they had observed. Their scientific efforts in astronomy and mathematics have mostly been lost, but the little information about Babylonian astronomy that has survived on clay tablet fragments show that it was the first successful attempt at describing astronomical phenomena mathematically. Some have said that all subsequent scientific astronomy in the West stemmed from this early Babylonian astronomy.

Around 600 BCE, Greek philosophers, included Pythagoras (570–495 BCE) and Socrates (470–399 BCE), tackled the questions about Nature's answers. Others followed, one of whom was Aristotle (384–322 BCE), who has been called the "father of Western philosophy." Greek mathematicians such as Euclid (mid-third century BCE) and Archimedes (287–212 BCE) made major contributions to the knowledge of science and mathematics. Archimedes was the first to rigorously estimate the number *pi*, and his calculational method[10] set the stage for calculus 2,000 years later. These early philosophers provided a foundation for Western scientific culture from their study of science and mathematics.

In the West, with the rise of the power of Christianity around 400 CE, religious philosophy took a dominant position over natural philosophy and science and replaced it with religious dogma as the gatekeeper of knowledge. However, by 1500, observations of Nature by Nicholas

---

[10] Steven Strogatz, *Infinite Powers: How Calculus Reveals the Secrets of the Universe*.

Copernicus, Johannes Kepler, Galileo Galilei, and others revealed obvious conflicts with the Christian dogma. Isaac Newton's 1687 book *Mathematical Principles of Natural Philosophy* is a book of science that described mathematically how gravity works on the movements of objects in the heavens as well as objects here on Earth. It is one of the most important books written on natural science.

The language of natural science is mathematics, and Newton developed a mathematical theory of gravity to answer questions about how the cosmos and the Earth work. Also, at this time, Francis Bacon described the logic[11] of the scientific method, which argued for knowledge to be based on observations of Nature. A hundred years later the philosopher David Hume's writings argued for the rejection of miracles and God as the designer of humans.

These scientists and philosophers worked to answer *how* the cosmos worked, while the answers to questions of purpose, or *why* things work, was left to churches and other philosophers pursuing metaphysics, a branch of philosophy that does not require mathematics. Those working on the *why* have offered many ideas but none of them have made contributions to scientific knowledge in the last 500 years.

In pursuit of their discoveries about Nature, scientist have faced uphill battles to gain acceptance of their theories from those with vested dogmas. Religious dogma from churches and secular dogma from old organizations have been hurdles to overcome for advancement of science. Some key examples of the conflict with church dogma will be covered later in more detail. At times, secular science departments have also acted to protect their favorite theories from new data. A famous example occurred when E. O. Wilson[12] in 1975 proposed an extension of Darwin's theory of natural selection by adding sociobiological factors; his new theory was originally strongly criticized by both scientific and religious organizations. At a science meeting, one scientist poured a glass of water on Wilson as a protest. Wilson took the view that criticism is the natural currency of progress and continued his research and let the results speak for themselves. Sociobiology is now a major field of scientific study.

Max Plank, founder of quantum physics, had a longer view of the problem of having new discoveries of Nature accepted:

> *A new scientific truth does not triumph by convincing its opponents and making them see the light, but rather*

---
11. Francis Bacon, *Novum Organum*, 1610.
12. Edward O. Wilson, *Sociobiology: The New Synthesis*, Belknap Press, 1975.

because its opponents eventually die, and a new generation grows up that is familiar with it.

Scientists continue with their research knowing that information comes by observing Nature, formulating supporting mathematical theories, and testing them. But a note of caution is necessary in trying to understand Nature. As Werner Heisenberg[13] noted:

[What] we observe is not Nature in itself, but Nature exposed to our method of questioning.

We do not know the depth or breadth of Nature's laws we observe. We have learned that existing theories may be rejected or replaced by new theories based on new data collected; hence, there remains uncertainty in our understanding of Nature with the theories we now have. Scientists continue doing their job with experiments and studies seeking to reduce the uncertainty in our picture of Nature.

Man's knowledge of Nature began early as our ancestors began to observe the world around them. Their eyes were their instruments for exploring the animals, plants, the weather, the stars, and geological features like volcanoes. They gazed at the sky and saw astronomical movements of the Sun, moon, stars, and occasional meteors. They used the information they gathered in their daily life as practical help. Their observations of the movement of the stars twinkling above were used in predicting seasonal patterns that aided hunting and later farming. Eventually, the discovery of metals—bronze and later iron—allowed man to make better observational tools as well as tools for many more of man's activities, including farming and waging war.

After writing was invented man started to record his observations, and systematic science began.

Around 500 BCE,[14] Greek philosophers offered theories about how the world worked and advanced theories of the structure of the heavens, but these were basically philosophical theses because there was no supporting experimental evidence. One example is Aristotle's theory that heavy objects fall faster than lighter ones. Aristotle also proposed a theory about the heavens: they were made of crystal spheres that carried the Sun,

---

13. Cathryn Carson, *Heisenberg in the Atomic Age: Science and the Public Sphere*, Cambridge, MA: Cambridge University Press, 2010.
14. "BCE" is used for dates before the beginning of the Common Era, which denoted as "CE."

moon, and stars eternally around the Earth[15] with unchanging circular motion. A hundred years later, Archimedes discovered the principle of buoyancy, which could be tested. He also made advances in mathematics by using the concept of infinity to calculate the number *pi*. Other major contributors were Euclid and Hipparchus around 300–200 BCE and Ptolemy around 100 CE.

As the Christian church grew larger and stronger over time, theologians attempted to assimilate the new scientific knowledge being discovered. However, they started with the thought that there was a supernatural all-powerful God and directed their prime efforts to tailoring everything as coming from their God. As mathematics and experiments were not being used by the church to understand Nature, but rather words from their God, their work remained only stories in the narrative.

But times change and in the Middle East, a more liberal religious environment emerged from 700 to 1200 in Bagdad, where thinkers of science and many different religions gathered and work together. Out of this mathematics flowered, with the invention of Arabic numbers, the zero, and algebra. These advances were later introduced to the Western world by the Italian mathematician Leonardo Fibonacci, who in 1202 completed *The Book of Calculations*, which popularized Arabic numerals in Europe.

The early theory of a geocentric universe adopted by the church lasted until Nicolaus Copernicus[16] proposed a heliocentric theory for the structure of the universe in 1567. Copernicus, a Polish clergyman and astronomer, published his theory that the Sun, not Earth, was the center of the universe. At that time, the church held not only that the Earth was at the center of the universe but also that the universe did not extend beyond our solar system.

That view was to change in 1608 when Galileo observed the heavens with his little homemade telescope and concluded that Copernicus's theory of a heliocentric system was correct and that there are moons around Jupiter.

Other scientists, like Tycho Brahe, made detailed observations of the movement of planets without a telescope. The data on the orbit of Mars that Brahe collected were sufficiently accurate that they were used as the

---

15. Although his science theories were incorrect, his philosophy was honored. He and Plato are considered the fathers of Western philosophy.
16. Dava Sobel, *A More Perfect Heaven: How Copernicus Revolutionized the Cosmos*, Bloomsbury, 2011.

source data for Johannes Kepler to discover mathematical laws (1619) explaining the variation in the speed of planetary orbits based on orbits being elliptical rather than circular with epicycles. Kepler's theories improved upon, but basically agreed with, the heliocentric theory (Copernicus believed orbits were circular). Then in 1687 Isaac Newton showed that Kepler's laws were correct as part of his own discoveries of the mathematical laws of motion and universal gravitation.

Galileo, a scientist with many interests, experimentally studied gravity by rolling objects down incline planes and dropping objects from towers. In one famous experiment Galileo dropped weights from the leaning tower of Pisa to show that objects with different weights fall at the same rate. An interesting historical footnote to gravity experiments was the experiment conducted by an Apollo 11 astronaut on the moon who dropped a hammer and a feather and demonstrated that Galileo was correct (they fell at the same rate from the moon's gravity).

Copernicus's heliocentric theory gave scientists an understanding of Earth's and the Sun's positions in space at roughly the same time that sailors began to sail around Earth and report that it was a sphere with oceans, tides, and continents. The time of the old thinking about the flat Earth and the Earth's position in the heavens was coming to an end.

The importance of mathematics as the language of Nature cannot be overstated. Our understanding of the universe is described by mathematical equations in physics, astronomy, chemistry, and other fields. Mathematical theories are used to predict phenomena before they are confirmed by observations. Examples: On the small scale, a nuclear particle, the positron, was suggested by a mathematical equation formulated by Paul Dirac before it was experimentally verified, and the neutrino was proposed by Wolfgang Pauli ten years before it was detected. On the large scale, the existence of the planet Neptune was predicted mathematically on the basis of its impact on the orbit of Uranus before it was observed by astronomers, and the existence of black holes was verified by gravitational pull on the orbits of nearby orbiting stars and by photographing the radiation of mass falling into one.

The English mathematician and first computer programmer Ada Lovelace summarized the importance of mathematics[17] in 1843 to describe Nature:

---

17. Claire Cain Miller, "Ada Lovelace" (obituary), *New York Times*, March 2018,

Math constitutes the language through which alone we can adequately express the great facts of the natural world.

Today, scientific discoveries are giving us an understanding of Nature at an accelerating rate. Over the years, technological improvements have vastly expanded the gathering of data from instruments for scientific exploration—new and larger telescopes with broader access to electromagnetic spectrum, more powerful microscopes, faster atomic accelerators, and so on. These tools have allowed research to be accelerated in many fields and opened new fields of study, and the improvement in computers and digital storage has broadened the capability to gather and analyze vast amounts of data from which we can obtain insights into Nature.[18] Below, several examples of advances in the disciplines of astronomy, physics, and biology are outlined to illustrate the discoveries of Nature's laws by scientists.

Discoveries have been made in many fields of science; however, only three (astronomy, physics, and biology) are highlighted for discussion here as examples of the evolution of scientific knowledge. Apologies are extended for not including the evolution of mathematics, chemistry, geology, and the many other disciplines used by scientists to gather our knowledge base of the universe; including them would have made this little summary book far too long. Sadly, that includes exciting new fields, such as evolutionary chemistry, epigenetics, and others.

The scientific search for the trajectory of Nature's universe is described in three stages:

- The creation of the universe by the big bang singularity and its initial ultrafast expansion or inflation and the continuing expansion and subsequent cooling that gave the forces and elements from which the stars and planets, including Earth, were formed
- The creation and evolution of biological life on Earth
- The evolution of the intellectual capacity of the *Homo sapiens* species to construct supernatural god concepts

---

18. Understanding gravity, a fundamental force of Nature, is an example. Some properties of gravity were noted and tested by Galileo. Newton described gravity as an attraction between masses. Einstein's theory of relativity saw gravity as a change in the curvatures of space–time caused by mass. Possibly in the future, gravitational theory will be expanded to include quantum theory with gravitons, yet-to-be-proven elementary particles that mediate the force of gravity.

This overview gives example insights into key scientific discoveries, scientists, and events in Nature's trajectory from the big bang to man's invention of gods during Nature's extraordinary 13.72-billion-year trajectory of the universe.

Many scientific tools have been invented to expand man's technological capabilities to perform the observations, experiments, and data analyses that are the basis for proving theories that give us an understanding of Nature. Several of the supporting technologies that have given man the critical tools to accomplish the research are outlined.

Technology

Stone tools and the ability to control fire were technological advances by our early ancestors that helped the hominin branch of the evolutionary tree survive. Around 3.3 million years ago—more than 2.5 million years before the genus *Homo*—our early ancestors invented primitive stone tools. Over time, the sophistication of stone tools improved, giving our ancestors new capabilities for accomplishing tasks that they could not easily perform. One such tool had stone tips added to wooden shafts to form spears, weapons that allowed them to kill prey at a distance. This gave them better safety in hunting large animals.

Our ancestors observed Nature and noted the changing of the seasons, weather patterns, and animal life behavior and put this knowledge to use in the invention of farming and domestication of livestock about 13000 BCE. Tools continued to be invented, including stone chisels, wooden mallets, stone hammers, and the ramp and lever for use in construction with heavy stones. With these simple tools the ancient Sumerians and Egyptians were able to construct sizeable structures, such as their ziggurats and pyramids.

In about 3500 BCE, man learned to make bronze by melting copper and alloying it with tin. This technological advance with metals ushered in the Bronze Age and the ability to make metal tools of greater utility than those made of stone or bone. About 2,000 years later, the Iron Age arrived and iron tools that had even greater strength and utility were made. As ironworking became cheaper, the use of iron tools rapidly proliferated throughout the Western world by about 1000 BCE.

In parallel, the ability to communicate improved, beginning with simple scratches on clay tokens first used in Mesopotamia around 7000 BCE for accounting purposes. The writing on clay tablets evolved and by

4000 BCE writing[19] in the form of cuneiform script became a general-purpose writing system on clay tablets in Mesopotamia. Writing became widespread in the Middle East and made available information that assisted in the growth of the first city-states. State laws and religious laws and narratives, which had long been transmitted orally, were now recorded and available to larger numbers of people. Writing gave rulers a new way to communicate and spread the king's laws and religious laws. Later, other societies developed their own writing systems, such as hieroglyphs in Egypt, and around 1200 BCE, the oldest alphabet evolved in Phoenicia; it was recorded on papyrus, a material easier to transport than clay tablets. With papyrus and later paper, the written word of scribes could be distributed readily.

The dissemination of scientific knowledge and tools used to study Nature got a boost in 1440 with the invention of the printing press by Gutenberg in Germany. For the first time, a cheap means of mass communication of information not emanating from religious scribes was available for the masses. Throughout much of Europe, the Catholic Church had monopolized knowledge using manuscripts handwritten in Latin by scribes, which were largely written and saved in monasteries. The Gutenberg press not only lowered the cost of information with its printed pages but, by using vernacular languages, also made scientific information accessible to more people outside of the control of the church.

Printing was an instantaneous success in light of the time. Within fifty years, most religious scribes lost their jobs and the Catholic Church's millennia-old monopoly over the written word was broken. Printing greatly improved the dissemination not only of scientific knowledge but also of that of other fields, including religion and governance. Copies of Martin Luther's sermons printed in 1517 spread widely and provided cheap fuel that dramatically ignited the Reformation and the following religious wars that altered the powers of governments throughout Europe.

By 1600, over twenty million books had been printed with this printing technology. Information about science experiments, such as reports on Galileo's observations of the heavens with his telescope, were made available to many.

Today the success of printing has been largely overtaken by progress in electronic data acquisition, analysis, and transmission. Vast amounts of

---

19. James C. Scott, *Against the Grain: A Deep History of the Earliest States* (New Haven, CT: Yale University Press, 2017).

data from scientific experiments are distributed, in many cases in real time, between facilities thousands of miles apart. Computing power, with its speed and the capacity of electronic networks, is an important tool used to gather, analyze, and distribute scientific data. Further, computing speed and capabilities now allow mathematical programs to model Nature's complicated actions, making computing an adjunct to and at times a substitute for physical experimentation by scientists.

In biology, the invention of optical microscopes in 1670 made direct observations of cellular life, such as microbes, possible for the first time. In 1931, the electron microscope was invented; ten thousand times more powerful than optical ones, it allowed for observations of viruses and phages and their interactions in real time. The decoding of DNA, the language of life, was greatly advanced in 1986 with the introduction of DNA sequencers. This technological advance allowed the first successful mapping of a human genome, the genetic material of an organism, which consists of RNA and DNA.

This program was launched in 1990 and completed in 2003 by a private company and government-funded programs at a cost of about thirteen billion dollars. Since then, sequencing costs have been greatly reduced. The cost of sequencing a genome is now measured in hundreds of dollars, giving a great boost to biological research.[20] Of course, none of this could have been accomplished without the technological advances that have occurred in computer power and the drastic lowering of the costs of computers and storage devices.

The rapidly expanding field of artificial intelligence (AI) is increasing man's ability to construct mental images and programs that extend his mental capabilities. One research area is the detailed functioning of the brain, where, possibly, the evolution of man's ability to conceive of gods and other supernatural spirits can be studied.

Many disciplines have contributed to our understanding of Nature. An overview of scientific progress in three disciplines will exemplify our increased understanding of Nature, the evolution of biological life, and man's capabilities to conceive of gods:

- *Astronomy* describes what is out there in the universe and our position relative to it.
- *Physics* describes the laws behind what is out there in the universe and inside of atoms and how they work.

---

20. Carl Zimmer, *David Reich Unearths Human History Etched in Bone, New York Times*, Mar. 20, 2018.

- *Biology* describes the evolution of life, *Homo sapiens*, and their brains with the mental capacity to construct religious narratives with gods.

Astronomy

Astronomy is one of the oldest of the natural sciences. Almost every tribe and emerging civilization observed the movement of the stars and planets, which were sources of both spiritual awe and utility for man. Many theories have been proposed about the universe, how the stars move, and their meaning to the observer. The methodical movements of the moon and planets and their positions in relation to stars were the basis for early celestial navigation and calendar settings for many things, such as when to plant crops and when to harvest.

Ptolemy, an astronomer and mathematician, wrote several scientific books around 150 CE[21] summarizing known science: astronomy, mathematics, and the geography of the Greco-Roman world. His thesis on astronomy presented a model of the universe as a set of nested spheres with Earth in the center of the universe. This geocentric view of the universe was held by the Catholic Church as dogma for over a thousand years.

That view of the cosmos changed when scientists made new observations that produced more accurate data about the movement of the planets, from which theorists were able to conceive the heliocentric theory. In 1567, Nicolaus Copernicus, a Polish clergyman and astronomer, published his theory that the Sun, not Earth, was the center of our solar system. Tycho Brahe made naked-eye observations of planets that gave data that were five times better than what was available, and with those data Johannes Kepler was able to conceive three mathematical theories on planetary motion.

In 1608, Galileo Galilei made observations of the heavens with the little telescope he had built and observed that some celestial bodies were orbiting other bodies that did not orbit Earth. One of Galileo's early observations centered on the movement of four objects moving around a brighter object in the sky that was later to be known as Jupiter. Figure 1 is a copy of the simple notations in Galileo's workbook made one night in 1608 when he recorded the positions of a bright heavenly body (Jupiter)

---

21. The term "Common Era" has its roots in a 1615 book by Johannes Kepler, who referred (in Latin) to the period since the birth of Christ as the "vulgar era," a term that evolved to "Common Era."

and four objects (Jupiter's moons[22]) that changed positions around it nightly. He concluded after many nights of observations that they were orbiting around the bright object; that is, the object had moons. Galileo's simple scientific observations (Figure 1) had profound consequences, for they were proof that all celestial bodies were not revolving around Earth. Some were revolving around other heavenly bodies. Galileo published his discovery in *Sidereus nuncios* (*The Starry Messenger*) in 1610. His observations were evidence supporting Copernicus's heliocentric model of Earth circling the Sun.

Galileo's observations conflicted with the established church geocentric dogma. In 1632 he published his views in *Dialogue Concerning the Two Chief World Systems*, which appeared to attack Pope Urban VIII. The Vatican refused to accept his conclusions, accused him of blasphemy, and brought him to trial; he was found guilty by the church in 1633. It took 202 years for the Vatican,[23] in 1835, to formally recognize Galileo's observations on celestial bodies as correct and to discard its dogma based on the geocentric universe theory.

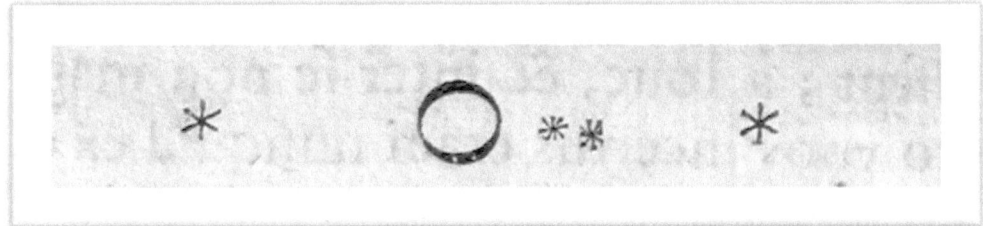

Figure 1 Galileo's Observations of Heavenly Bodies (Jupiter and Four Moons)

The authority of God's word on the heavens was shown to be inferior to Nature's authority based on physical observations. With these and other experimental observations, the scientific community has discovered that moons orbit not only Jupiter but also other planets.

Sixty years later, Isaac Newton in England formulated[24] a mathematical theory of motion that described the movements of the

---

22. Since that time many observations have been made of the orbits of Jupiter's moons (there are seventy-four other moons in addition to the four Galileo observed), and spacecraft have been sent there to observe several of them at short range.
23. William D. Montalbano, *Vatican Finds Galileo 'Not Guilty,'* William D. Montalbano, Washington Post, Nov. 1, 1992,
24. Isaac Newton, *Mathematical Principles of Natural Philosophy*, first published July 5, 1687.

heavenly bodies, including planets and moons. He noted that the forces controlling planets were the same forces that were controlling bodies on Earth, such as apples falling from trees and cannonballs fired from guns. Newton's theory on gravity most people know about. It supports the universality of Nature's laws. Again, a science theory based on evidence conflicted with the church's dogma that declared that physical laws in the heavens were different from those on Earth.

Enlightenment from scientific observations extended to the Americas when in 1752 Benjamin Franklin demonstrated that electrical discharges in the heavens (lightning) could be collected with kites flown in electrical storms and that these discharges had the same properties as the electrical discharges he could generate on the ground. Franklin, like Newton, was observing the universality of Nature's laws; they are the same in the heavens as they are on Earth.

Galileo's telescope was just the beginning of telescopes as observational tools; the construction of larger optical telescopes that followed has allowed for deeper studies of distant objects in the cosmos. Figure 2 lists some of the optical telescopes that have been built: large ones on Earth, such as the Keck, and one, the Hubble telescope, that operates in space.

Newton's theory of gravity and supporting observations by many astronomers began to open the door to understanding Nature's universe by showing that the same laws that control our little planet also control all of the universe. But understanding gravity, one of the four fundamental forces of Nature, has turned out to be more interesting than Newton could have dreamed. Newton described gravity as an attraction between masses, and time was not considered in his theory. Einstein saw it differently, and in his general theory of relativity he included time in his equation, which described gravity as a change in the curvatures of space–time. This equation gives a deeper insight into Nature, which we now know allows space–time to be curved and warped and to have ripples.

As telescopes grew larger and more powerful, some of the mysteries of the universe became a little more understandable through observation; for one, the universe and objects in it are older, larger, and more interesting than previously thought. In 1923, Edwin Hubble, analyzing data on the movement of stars he had obtained with the 100-inch Hooker telescope, determined that there are other galaxies beyond our own Milky Way galaxy. In 1929, his observations showed that these faraway galaxies are all speeding away from us as if from a single point in space. From these data, the physicist and Catholic priest Georges Lemaitre theorized that there was a big bang from a single point that started the creation and outward expansion of galaxies, and hence the universe. His calculations followed from Einstein's general theory of relativity and became known as the big bang theory. Though he was a priest, Lemaitre insisted that his thoughts were strictly scientific. When Pope Pius XII argued in 1951 that the big bang was scientific evidence for God, Lemaitre dissuaded the Pope from further pronouncements on this view, for God was not included in the calculations.

Today the light-gathering power of very large telescopes, such as the 394-inch Keck telescope and the 94-inch space-based Hubble telescope, is being used to observe galaxies formed only a few hundred million years after the big bang. It is expected that soon, a new space telescope, the Webb telescope, will be launched as a replacement for the Hubble and provide a larger 250-inch light-gathering capability to look deeper in the universe. Figure 2 lists the diameter of the light-gathering surface of telescopes, from the first telescope made by Galileo to telescopes with a diameter of 394 inches by the end of the twentieth century.

	Lens Diameter (inches)	Operational Date
Galileo	0.62	1609
Herschel	50	1815
Hooker	100	1917
Palomar	200	1948
Hubble *	94	1990
Keck 1 and 2	394	1995
Webb* **	250	2018

Figure 2 Growth of Optical Telescopes
(*space-based, **to be launched in 2021)

By the beginning of the twentieth century, there were two leading theories of the universe purporting to explain the expanding universe observed by Hubble in 1929. One was the new big bang universe theory that posited a common hot beginning for all matter, followed by the expansion and cooling period we are in now. The other was the steady-state universe theory, which posited a continuous and steady creation and death of matter and a universe with no beginning and no end. Today the space-based Hubble telescope, other very large ground-based telescopes, and radio telescopes enable us to see beyond our galaxy and observe other galaxies billions of light-years away. From these telescopes, especially the Hubble's deep-space observations, scientists have obtained images of distant galaxies, some of which existed when the universe was less than 2 billion years old. These distant galaxies in our expanding universe provide a time machine for determining the age of distant objects existing—some of them—almost at the beginning of the universe 13.72 billion years ago.

An accidental discovery of the existence of cosmic microwave background (CMB) radiation with a radio telescope in 1964 gave some insight into the structure of the early universe. The discovery of the CMB radiation led most scientists to eliminate the steady-state theory from consideration and adopt the big bang, which could predict the temperature of the radiation data from the CMB. Figure 3 is a CMB image of the surface of the universe at 400,000 years after the big bang showing the variations in distribution of matter at an early date in the expansion of the universe. Since the discovery of CMB, additional observations with more powerful tools looking at the CMB radiation surface with the Boomerang experiment in 1997 and the WMAP (Wilkerson Microwave Anisotropy Project) in 2001 confirm the big bang theory. The steady-state theory of the universe is an example of a beautiful theory that has been rejected because it did not comport with the experimental data.

Figure 3 Cosmic Microwave Background Radiation

Beyond optical telescopes, other types of telescopes and instruments have been constructed to explore the heavens. Ones have been built to study different frequencies of the electromagnetic spectrum, including infrared, radio, X-ray, and gamma rays. One successful observation was made by an array of eight radio telescopes that together act as a giant virtual telescope that was synchronized to focus on the same object in space at the same time. This novel experiment[25] by an international network of radio telescopes, called the Event Horizon Telescope (EHT), was able to photograph an image of a black hole and its surrounding

---

25. Ota Lutz, *How Scientists Captured the First Image of a Black Hole*, JPL, California Institute of Technology, April 19, 2019,

material (Figure 4). Although the black hole is indeed black with no radiation escaping, the accelerated matter swirling around and into the black hole, powered by its tremendous gravitational force, does radiate, and it is this halo of radiation around the black hole that gives us the image of the supermassive black hole. This black hole has a very large mass of about seven billion times that of the Sun's and is located at the core of the supergiant elliptical galaxy Messier 87.

Other objects of interest in exploration of space include neutrinos and gravitational waves. For detecting neutrinos, several particle detectors, such as the Super-Kamiokande in Japan, have been built. Since neutrinos are very difficult to observe (they are electrically neutral), this detector is

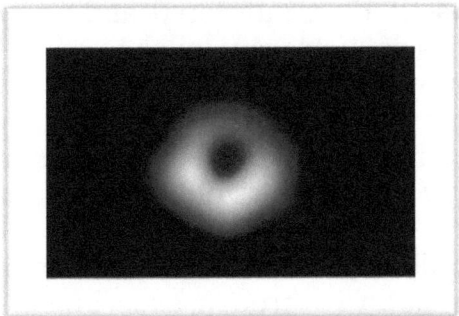

Figure 4 Black Hole Image

massive and contains fifty thousand tons of water. Scientists announced that another neutrino detector, the Ice Cube Neutrino Observatory, has traced extremely high-energy neutrinos that hit their Antarctica-based research station back to their point of origin, a blazar[26] located 3.7 billion light-years away in the direction of the constellation Orion. This was the first time that a neutrino detector had been used to locate an object in space.

Successful observations have also been made by the Laser Interferometer Gravitational-Wave Observatories (LIGO, two in the United States and one in Italy), which have detected gravitational waves[27] or ripples. This view of gravity has been observed by instruments

---

26. A blazar is an active galactic nucleus (AGN) with a relativistic jet (a jet composed of ionized matter traveling at nearly the speed of light) directed very nearly toward Earth.
27. Davide Castelvecchi and Alexandra Witze, "Einstein's Gravitational Waves Found at Last," Nature, February 2016. The gravitational waves of two black holes (with masses of 29 and 36 solar masses) merging around 1.3 billion light years away were detected. The mass of the new black hole resulting from the merger of the two was 62 solar masses. Energy equivalent to 3 solar masses was emitted as gravitational waves.

measuring the ripples in space–time from the collision of massive black holes. In both 2015 and 2016, collisions of massive black holes more than a billion light-years away were detected from the ripples in space–time and recorded. The ripples from another collision, this one between two neutron stars, was detected in 2017 and confirmed by optical telescopes. These observations have further verified Einstein's general theory of relativity.

Recent attempts to get data very shortly after the big bang have not succeeded. A recent example was an experiment to detect the primordial gravitational waves fractions of a second after the early expansion, which would give data for the earliest glimpse after the big bang. However, impacts of intergalactic dust interfered and warped the original data. With these new data scientists were quick to admit that the results of the experiment were not valid.[28]

From experiments over the last five hundred years, our vision of an Earth-centered universe has changed to one of a dramatically expanding universe that is 13.72 billion years old in which our Solar System is but one star in a galaxy, the Milky Way, with a 100 billion stars in a vast universe with 400 billion galaxies. Our understanding of the universe has come by observing Nature—through experiments—as its causal agent. Nature's universe requires no supernatural designer, for what we observe are products of Nature's laws doing their job.

Astronomy and mathematics provide the base for our look into the cosmos. Freeman Dyson[29] noted that studying the cosmos is like visiting "a zoo full of wonderful creatures." He added,

> *Most of the mathematicians are busy admiring the architecture, while the physicists are admiring the animals. Which is more important isn't, to me, the interesting question. The interesting question is, why do they fit together so well?*

But mathematics is not used only in astronomy; it is the language of all of science.

---

28. Brian Keating, *Losing the Nobel Prize*, W. W. Norton & Company, 2018.
29. Siobhan Roberts, *Celebrating the Eclipse That Let Einstein Shine*, New York Times, May 28, 2019.

Physics

For a long time, philosophers composed theories that included everything. Some early philosophers gave us insights into Nature, but mainly they got much of the science wrong, such as Aristotle on gravity—heavy bodies fall faster than lighter ones. Some had better hypotheses about Nature; for example, the Roman philosopher Lucretius in 57 BCE stated,

> *Everything is made of invisible particles [atoms], which are eternal.*

This is a helpful insight, partially true, but there are also many particles that spontaneously disappear by decaying into others and others that zip in and out of existence. Further, there may be no particles at all, just—probably—waves. But nobody was around to test the theories and provide evidence, for the early philosophers had a handicap: there were no experiments to support their insights.

About fifteen hundred years later, physicists (for example, Galileo Galilei and Isaac Newton) made observations and experiments on the *invisible force* of gravity, and another five hundred years later, there were experiments by nuclear physicists on *indivisible particles and waves* to support their theories.

Galileo in 1608 built a little telescope that opened up the heavens for observation and Isaac Newton formulated a theory of gravity in 1686 that told how objects, such as planets and cannon balls, move. There have been many other discoveries, but we will focus on a few examples of the scientists and their advances in three fields, astronomy, nuclear physics, and biology.

In 1865 a physicist attacked the mystery of magnetism, which had been used by magnetic compasses since the eleventh century. But how did a magnetic compass work? This question was answered by the Scottish physicist James Clerk Maxwell in 1865 when he published his theory[30] of electromagnetic radiation, which describes light being composed of electric and magnetic fields. His mathematical equations described the generation of electromagnetic radiations with various wavelengths, from radio waves to gamma waves. Maxwell's equations are still being used to calculate the working of electromagnetic gadgets.

---

30. James Clerk Maxwell, *A Dynamical Theory of the Electromagnetic Field*, 1865.

As late as the nineteenth century, a leading physicist, Lord Kelvin, using the scientific knowledge of the time, calculated that our Sun as a source of energy would be able to shine with its apparent brightness for only tens of millions of years. However, this position conflicted with new geologic data being gathered suggesting that Earth is much older, possibly ten to a hundred times older. Also, late in the century, physicists were having problems describing mysterious radiations coming from certain natural materials, such as ores that contained the element radium, but also from vacuum tubes that were being made at the time.

These problems were solved when physicists learned more about nuclear energy. The first step in solving the problem of radiations from radium began with discoveries from X-ray experiments made by the German scientist Wilhelm Röntgen in 1895. Henri Becquerel, in France, first observed radiations coming from uranium salts in 1896. This was followed by Marie and Pierre Curie's experiments on natural radioactivity from uranium salts and other elements, such as thorium and radium. The three won the Nobel Prize in Physics for their research on radioactivity in 1903.

In 1897, British physicist J. J. Thomson showed that the cathode rays in vacuum tubes were from an electrical discharge and were composed of a previously unknown negatively charged particle, the electron. In other experiments, other radiations were being discovered by Ernest Rutherford: the alpha particle (two neutrons and two protons), the beta particle (an electron), and gamma rays (high-energy electromagnetic radiations).

Max Planck in Germany, while studying blackbody radiations in 1900, offered a simple theory that wave-like radiated energy, such as heat and light, could be emitted only in small quantized amounts—quanta—whose energy depended on the frequency of the radiations, as noted in his simple formula $E = hv$ (h is a constant and v is the wave frequency). This was the conceptual beginning of quantum mechanics. Others picked up the quanta concept and advanced our understanding of it. Albert Einstein published a paper on the photoelectric effect (1905) that used the quantum concept. This work won him the Nobel Prize in Physics.

It is hard now to believe that at the beginning of the twentieth century, physicists did not know what an atom looked like. Theories on the structure of the atom begin appearing in 1902. Others followed and by 1911, Ernst Rutherford[31] had evolved a theory (revised by Niels Bohr in

---

[31] Ernst Rutherford, *Radio-activity*, 1904.

1913) that gave a picture of a central nucleus composed of protons and neutrons surrounded by a cloud of electrons in discrete and stable orbits that obeyed quantum rules. Although details have been revised by later work on quantum mechanics, this is the general picture still taught in schools. Bohr's addition of quantum mechanics into the explanation of how atoms are constructed made quantum mechanics a major field of physics.

Albert Einstein also published in 1905 his special theory of relativity, which introduced the mass–energy equivalence formula, $E = mc^2$ (m is mass and c is the speed of light). In 1915, he published the general theory of relativity, which revolutionized our understanding of gravity and the motion of matter in the universe through the introduction of the concept of space–time. In 1919, Englishman Arthur Eddington led an expedition to South America to test Einstein's theory of relativity during a solar eclipse. The experiment observed the bending of light from stars when they are passing by a large mass, such as our Sun. The exact deflection of starlight caused by the curvature of time–space from the Sun predicted by Einstein's theory was measured and found correct, thus verifying his general relativity theory. Einstein's general theory of relativity provided a mathematical theory of gravity that gave cosmology a base for calculations.

Quantum mechanics theories give views of Nature on a small scale that are seemingly counter to the large-scale world we know. For example, particles can act like waves and waves can act like particles. The light we observe is made of photons, which can be both particles and waves. Your eyes are activated by certain photon wavelengths, yet light can collide with matter, scatter, and be measured as particles. Nature described by quantum theories is different from the world described by classical physics.

The quantum mechanical view of Nature can be very strange to us, but experiments have proven many of the quantum mechanics theories. From the beginning, physicists have been unsettled by quantum theory, such as entanglement of quantum states in atoms. In 1927, Einstein argued that the quantum mechanics theory of particle entanglement that he called "spooky action at a distance" was incomplete but correct.[32] Although the quantum entanglement theory is a major disparity between classical and quantum physics, it has been supported by experiments that indicate that

---

[32] *Einstein Attacks Quantum Theory*, New York Times, May 4, 1935.

it is a real description of Nature. Together, classical and quantum physics form our incomplete description of Nature.

In 1928, the British astronomer Fred Hoyle and others formulated a theory of the universe called the steady-state theory and argued that it was competition to the single-source (big bang) theory proposed by others. A year later, he coined the term "the big bang," intending it as a pejorative for he thought it only followed the religious description in Genesis with god making the universe with a bang. However, the 1964 the discovery of cosmic background radiation removed support for the steady-state theory from scientists in favor of the "big bang theory." The latter continues to survive many challenges.

The big bang theory holds that after initiation, there was brief moment (10^{-43} seconds) of hyper-expansion of the universe. Over the 13.72 billion years of expansion, the universe is estimated to contain one hundred billion galaxies, each having a hundred billion stars—staggering numbers for sure. The big bang is the source of everything in the universe.

Our understanding of the forces and matter that make up the universe created by the big bang began to emerge in the 1930s from scientific experiments. The bombardment of elements with nuclear particles achieved unexpected results, such as turning one element into a different element with a different atomic number, raising new questions.

Experiments in 1938 in Germany using neutrons to bombard heavy elements split the atom and showed the release of large energy. Experiments demonstrating the fission process splitting uranium atoms were conducted by Otto Hahn, Fritz Strassmann, Lise Meitner, and Otto Robert Frisch. In fission, the absorption of a neutron causes the uranium atoms to split into several smaller atoms, plus the release of energy as a small amount of the uranium mass is converted into energy that push the fragments apart. The energy of mass had been predicted by Einstein's famous $E = mc^2$ theory.

Further, with other neutron bombardment experiments, scientists have been able to make other elements that fission, such as plutonium-239, which is made from the absorption of a neutron by uranium-238.

Within seven years, the fissioning of uranium-235 was employed as the energy source in the first atomic bombs ("Trinity" at a New Mexico test site and "Little Boy," which was dropped on Hiroshima, Japan, in 1945). A few days later plutonium-239 was the fission material used in the second atomic bomb dropped on Nagasaki, Japan.

The problem of the vast amounts of energy from the Sun that Lord Kelvin was struggling to explain was solved by Hans Bethe in 1939 with the theory that nuclear fusion of light elements of hydrogen at the center of the sun release the enormous energy from the Sun. He calculated that the fusion of hydrogen in the super-high density and temperature at the center of the Sun could supply the energy needed to power the Sun for billions of years, thereby solving the 1920 problem of the Sun's life.

The fusion process was later used in hydrogen bombs in 1952. A hydrogen bomb uses a fission bomb and adds light elements (hydrogen and lithium isotopes) for nuclear fusion, giving additional explosive energy. The hydrogen bomb's power is scalable to thousands of time greater than the first atomic bomb explosion (20 kilotons). Russia has tested a hydrogen bomb with a release of 50 megatons of energy.

I was a witness to a test explosion of a 20-kiloton atomic bomb. The sun-like brightness and radiative power of the explosion over the Nevada desert test site was awesome, for I could not only see the explosion but feel the heat from the blast. It was particularity poignant knowing that the explosion had come from two half-orange-sized Pu-239 hemispheres that were quickly compressed by the bomb's high-explosive shell and exploded with power a million times greater than a comparable chemical explosion.

The hemispheres in the bomb that the author saw explode were comparable to two test hemispheres that his laboratory group had tested for their nuclear properties (criticality[33] test) at Los Alamos. Even in the laboratory, one could literally feel that the two little metal hemispheres were different, for when handled alone in air during test preparation they produced enough internal heat (15 watts) to cause their temperature to slowly rise. Although the hemispheres were mainly Pu-239, the heat was generated mostly by the spontaneous fission from a small impurity, Pu-240. The Pu-240 was an unwanted side product made in nuclear reactors while making the Pu-239 when some of the Pu-239 captured a neutron instead of fissioning.

Earlier two scientists had been killed in the lab in a mishandling accident with Pu-239 cores in earlier experiments on assembling cores to be used in weapons. The critical mass (amount of material to initiate an explosion) of Pu-239 is less than twelve pounds when neutron reflectors are used. One is very, very careful and respectful when handling these little hemispheres in the lab.

A second awesome sight I saw somewhat later was an Apollo launch. Watching a six-and-a-half-million-pound Saturn rocket lifted off with seven-and-a-half million pounds of thrust, creating a mighty roar and

---

33. The point at which a self-sustained nuclear fission reaction occurs.

accelerating the Apollo 8 capsule on its mission to the moon, was indeed awesome. This scene of rocket power unfolded according to the simple equation $F = ma$ (force equals mass times acceleration).

The two awesome events I have experienced happened as one would have predicted from the mathematical laws of Nature. The feeling of awe is a universal human feeling that has been felt by many—the religious, the spiritual, and the atheistic—when they have experienced certain events that trigger the brain's "awesome feeling button."

In 1949, Willard Libby developed the technique of using the radioactive carbon isotope (C-14)'s decay to date organic materials, such as old bones or plants. The use of the dating technique for fossils to older times has been extended by using various other radioisotopes with longer half-lives for dating other items. Carbon-14, with a half-life of 5,730 years, is useful for dating samples up to 50,000 years old. Other elements, for example, uranium-235, with a half-life of 703 million years, are used to date materials up to a few billion years old. These and other radioactive elements have been most useful for dating the ages not only of fossils but also other materials of interest to archeologists.

Many tools have been invented to assist in scientific experiments. In physics, one of the most successful tools has been the nuclear particle accelerator, an instrument that accelerates nuclear particles to very high speeds and smashes them together to study their interactions. The first accelerators were built in 1928, the Van de Graaff electrostatic accelerator by Robert Van de Graaff and the linear accelerator by Rolf Wideröe. These were followed by the invention of the cyclotron by Ernest Lawrence in 1930. Because of the cyclotron's scalability to very high energies, it is the primary accelerator used today. From the study of the collisions of the accelerated particles, scientists have been able to observe the building blocks of nuclear particles.

In the early days of the accelerator, experimental physics became known as "telephone physics" because the laboratory that first built and operated the most powerful accelerator could make new discoveries from the higher energies attained and beat out other investigators. To claim the credit for being first to discover new particles or their interactions it was necessary to quickly telephone the other laboratories before they had completed a comparable experiment. Paper reports would have taken too long.

The discovery of several new particles with these accelerators had theorists scrambling to explain them. Murray Gell-Mann and George Zweig theorized that some particles were composed of elementary particles, the quark, that, used in combinations of three, make the composite particles, the neutron and protons in all atoms.

The first cyclotron accelerator built in 1930 operated on a desktop in a university laboratory. Since then the race to build larger accelerators has led to increased size and power. The construction of the largest accelerator, the Large Hadron Collider (LHC), a circular accelerator with a seventeen-mile proton particle raceway, was completed in 2013. The LHC is the most powerful accelerator and the most expensive scientific tool ever built.

The very high-energy protons smashed together by the LHC have provided the high energies for new discoveries, and in 2016 the LHC was used to discover the elusive Higgs particle, which gives mass to all particles in Nature. This was a major discovery in particle physics and a boost to the Standard Model[34] of physics, which catalogs what we know about elementary particles and the forces with which the universe is constructed. However, the model does not include gravity, dark matter, and dark energy. Much work remains to put all of these into a cohesive theory.

Although the theories of physics have provided successful tools for many of today's applications in astronomy and engineering, scientists are seeking a way to make quantum theory and Einstein's general theory of relativity compatible in describing gravity. Richard Panek[35] noted:

> *We can say gravitation is one of the four fundamental forces, but it's such an outlier that the word "force" becomes nearly meaningless. The strong nuclear force (which keeps atomic nuclei intact) is about 100 times stronger than the electromagnetic force (which creates the light spectrum), which in turn is up to 10,000 times stronger than the weak nuclear force (which facilitates the subatomic interactions responsible for radioactive decay). Three forces, all within six orders of magnitude of one another. Then comes gravitation. It's about a million billion billion billion times weaker than the weak nuclear [force].*

We can calculate gravity's effects, but we clearly do not understand it fundamentally. It is possible that in the future the theory of gravity will be

---

34. Lisa Randall, *Knocking on Heaven's Door: How Physics and Scientific Thinking Illuminate the Universe and the Modern World*, New York: HarperCollins, 2011.
35. Richard Panek, *Everything You Thought You Knew About Gravity Is Wrong*, Washington Post, August 2, 2019.

modified to include gravitons, posited elementary particles that are thought to mediate the force of gravity in quantum calculations.

Other theories have been proposed to explain the structure of Nature. One is string theory, which places small vibrating strings[36] as the fundamental unit of particles and forces. So far, this theory has failed to calculate many of the basic constants of Nature. Nature remains elusive.

Although much has been discovered to date, there are still many unknowns. Unknowns in Nature to scientists are normal for they give focus to research efforts to understand the fundamentals of Nature's laws. Here are some examples of key physics theories under study:

- Big bang. The big bang was a single cosmic event, a quantum fluctuation of a fraction of a second, followed by an expansion of space–time with matter and forces appearing. Subsequent expansion and cooling of the universe continues.
- Antimatter. The big bang produced an equal amount of matter and antimatter, yet in an instant the antimatter was gone. Normal matter contributes 5 percent of the mass-energy of the universe.
- Dark matter. A substance responsive to the gravitational force that does not emit or interact with electromagnetic radiation, such as light. It contributes 27 percent of the mass-energy of the universe.
- Dark energy. A force in space–time that, after five billion years, has been observed to cause the universe to expand at an accelerating rate. It contributes 68 percent of the mass-energy of the universe.
- Black holes. Have been indirectly observed, but theories describing their internal structure are preliminary and do not conform to present theories.
- Gravity. Although theories of gravity in the large-scale world are known, gravity on the small (quantum) scale is not known. Something is missing in the fundamental understanding of gravity.

On a separate note, mathematics is the language that describes Nature, but the question remains, *why* does mathematics work as the language of Nature? Nature's reality is proving to be elusive to our reasoning powers, but mathematics allows us to learn and engage with Nature.

---

36. Brian Greene, *The Elegant Universe,* W. W. Norton & Company, New York, 1999.

IV. The Universe

Once upon a time, there was no time, no space, no matter, no Sun, no Earth—no universe. Then there was a big bang,[37] a creative event with properties not well understood, a singularity that created the universe and set the trajectory of the expansion of the universe with the forces, energy, matter, and space–time, as the physicists like to say. In a mere fraction of a second (10^{-32} second), the universe expanded (in a period called the inflationary epoch) by nearly as many orders of magnitude (10^{26}) as it would in the following 13.72 billion years. The quantum theories that give us an understanding of inflation are based on empty space having energy, for empty space is complicated, with virtual particles[38] that pop in and out of existence.

This burst of cosmic inflation has been followed by a much slower inertial expansion of the universe. About 6 billion years after the big bang, a force, dark energy, is causing an accelerating expansion of the universe that is observed today. The expanding universe has continued cooling from the extreme temperatures of the instant of the big bang to its present level of a few degrees (2.7 °K) above absolute zero.

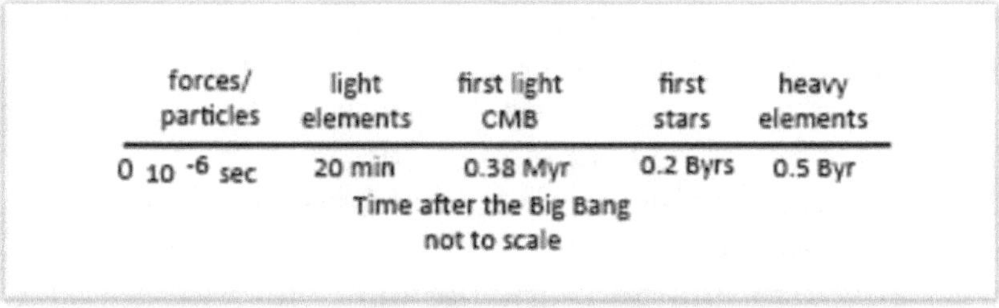

forces/particles	light elements	first light CMB	first stars	heavy elements
0 10^{-6} sec	20 min	0.38 Myr	0.2 Byrs	0.5 Byr

Time after the Big Bang
not to scale

Figure 5 Evolution of the Early Universe

From the very early moments, cooling of the inertial expansion allowed the elemental forces and matter to appear. The forces of gravity and the electromagnetic force are long-range forces whose effects can be seen directly in everyday life, while the strong and weak forces are short-range and govern nuclear interactions. Dark energy appears to be doubling the

---

37. There are other theories, but this one has the best supporting data.
38. Lawrence Krauss, *a universe from nothing*.

universe every 10 billion years. Both dark energy and dark matter are poorly understood at this time, but they probably emerged at the beginning. Within a few minutes, the cooling from the expansion gave the right conditions for the appearance of matter as subatomic quarks that would aggregate to produce the matter of atoms—protons, neutrons, and electrons. Creation of matter began with equal amounts of antimatter and matter being created, but a process not fully understood led to an imbalance of matter and antimatter. After twenty minutes, the particles began combining into the atomic nuclei of the light elements hydrogen (H) and lithium (Li), the building blocks of the universe. These light elements initially existed only in a plasma state because of the extreme temperature at that time.

As the universe continued to expand and cool, events began to occur more slowly as different phases of the state of the big bang's energy appeared. (An analogy to the universe's cooling phases is the phase changes with H_2O: from a high-temperature cloud of elemental H and O to the formation of H_2O molecules as steam (gas); then with further cooling, H_2O forms as a liquid (water), and finally, with further cooling, H_2O turns into a solid (ice).

The cooling of the light elements, hydrogen, helium, and lithium atoms, in the plasma state allowed atoms to capture electrons and form elements that in turn allowed radiation to escape from the plasma after about 400,000 years. This radiation has been observed and is known as cosmic microwave background radiation (CMB). Measurements of CMB radiation today giving a temperature of $2.7\,°K$ are consistent with calculations of the cooling from the expanding universe over the 13.72 billion years since the big bang. CMB radiation was first detected accidently in 1964 as an unexplained hiss in an antenna during experiments on terrestrial microwave communication that appeared to come from all directions.

After the first 100 million years, the light elements formed by the big bang had scattered across the cosmos and they began to coalesce in regions from the force of gravity. These lumps of matter gave rise to the first light-element stars. Over the next few hundred million years, these light-element stars—which are short-lived—formed, lived their lives, burned (through fusion) their light-element fuel, and produced in their super-dense interior stellar ovens heavier atoms, such as carbon, oxygen, and iron. Upon the deaths of these stars after a few hundred million years, they exploded in spectacular stellar explosions called supernovas and flung their newly made heavier elements into the cosmos. The dust of the new heavy elements, plus that of the light elements left from the big bang, provided the material for new heavy-element-type stars, like our Sun. Since then stars, light- and heavy-element ones, continue to form, live their lives burning the nuclear fuel, and, when it is consumed, explode.

The residuals from the death of a star depend on the mass of the star. The results of exploding low-mass stars can be a white dwarf, a neutron star, and other objects. The death of very large stars is different and results in a collapse into a black hole. This has been happening and continues to happen throughout the universe today. Figure 5 lists the sequence of cosmic events that led to the first stars and the appearance of heavy elements that we see today in heavy-element stars like our Sun.

What has been learned about Nature from scientific observations and experiments gives a general overview of the evolution of the universe. When it occurred 13.72 billion years ago, the big bang spawned the expansion of the universe and created the spectacular diversity of cosmic objects we see, measure, and study today. The universe continues to expand, but about 6 billion years ago the force of dark energy in the universe overcame the gravitational deacceleration force and the expansion rate increased, leading to the accelerating expansion of the universe we observe today.

Little is known of the roles dark matter and dark energy played in the early formation of the universe. From what is known, it is estimated that the universe is composed of 26 percent dark matter, 69 percent dark energy, and about 5 percent normal matter. This dynamic universe we know about today is far from the steady-state theory of the universe of a hundred years ago.

About 9.3 billon years after the big bang, our solar system—the Sun and several planets, one of which we call Earth—formed from the cosmic dust of past exploding stars. Our Sun is a moderate-size star and our Earth is a moderate-size planet in an orbit that gives it a moderate temperature. At our distance from the Sun, Earth has a "Goldilocks" environment for biological life, neither too hot nor too cold. On Earth, biological life evolved after another 4.5 billion years, also from the cosmic dust, the material of the Earth.

The creation of the universe, including our Earth and man and other biological life, occurred under Nature's laws and this has been accomplished without a designer—that is, without the involvement of a supernatural god. The sequence of some key events in the early evolution of the universe, Earth, life on Earth, and *Homo sapiens* as understood by scientists is given in Figure 6. It presents a picture of the dynamic unfolding of the universe from the big bang (time 0) with the creation of particles, stars, planets and galaxies, and life on Earth, including *Homo sapiens*.

The universe we observe today is staggeringly large as well as staggeringly mysterious. The universe has expanded from a small speck to a universe that is 91 billion light-years in diameter and that contains billions and billions of galaxies, stars, planets, and other celestial objects. What a challenge it is to make sense of it all! But that is what scientists have been trying to do—gather information from observations of Nature's laws and formulate theories to explain it all.

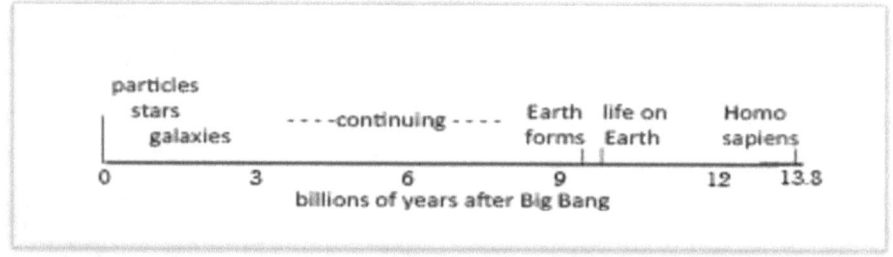

Figure 6 Evolution of Earth and Life

Also, on the smallest scale, information gathered from experiments is yielding insights about Nature and its laws that describe the basic building blocks of the elements, namely quarks, that make up the protons and neutrons and the four forces, including the glue that holds particles together and controls their interactions with other particles and forces.

Scientists have learned that an understanding of Nature is elusive. What we believe we have learned may not be the underlying reality of Nature and our existing theories may have to be revised, for there may be deeper laws to discover that we have yet to stumble upon. For example, scientists are still trying to understand why quantum theories that work well enough in describing the small world of atoms fail in the larger-scale case of stars, where Newton's and Einstein's theories of the large work well. However, albeit with gaps in our knowledge, scientists have been able to construct theories that link the existing data on the creation of the universe (heavens) after the big bang to the formation of Earth and the evolution of man.

These observations have given scientists an initial understanding of Nature's laws. Theories have also been inferred from data collected on objects that are unseen and outside of known theories, such as hypotheses describing dark matter, dark energy, and black holes and their interactions with known matter. But the fundamentals of the structure of the universe are not known, nor do we know whether it is finite or infinite. So, it is apparent that there is much work ahead before we have a full understanding of Nature's big bang and the resulting universe.

But we do know that on Earth, biological life has evolved and our species, *Homo sapiens,* has evolved the ability to not only observe the universe but also conceive of complicated images, such as supernatural gods. These two evolutionary events—evolution of biological life with *Homo sapiens* and the evolution of his capabilities for mental constructs, including stories with supernatural gods—are outlined in the *once upon a time* stories in the next two chapters.

V. Biological Life

Once upon a time a planet, our Earth, formed as part of our solar system 4.5 billion years ago. There was no biological life on this newly formed planet. With the Sun providing continuous heat to the Earth's moderate-temperature surface and with pools of chemical soups made from the cosmic dust on Earth, the stage was set for the evolution of life. It didn't happen quickly, but after more than 700 million years the first self-replicating molecule evolved on Earth. Once life evolved on Earth it proliferated, and after 3.8 million years of evolution of many generations of various species, from single cells to vertebrates, and after another 600 million years, the vertebrates *Homo sapiens* appeared.

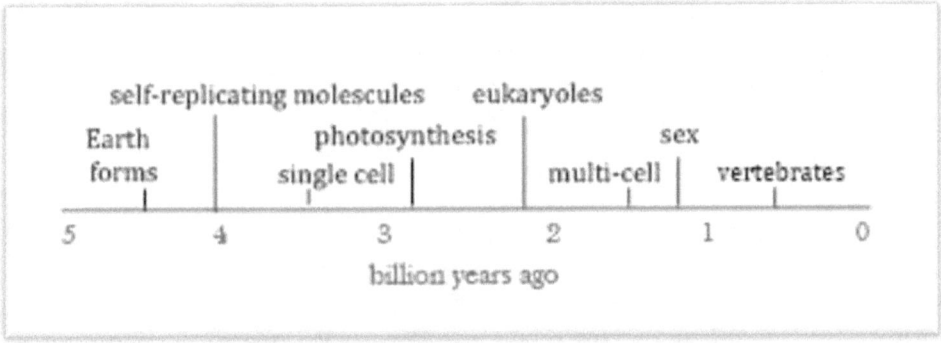

Figure 7 Evolution of Life on Earth

Carl Sagan, a famous astronomer, has noted that

> *man is composed of stardust.*

The meaning of Sagan's comment is that life evolved from the dust of stars that formed our solar system, including the planet Earth. This dust from which life as we know it sprang consists of both light and heavy elements. The light elements came directly from the big bang, while the heavy elements came into being in the core (a fusion furnace) of stars, which upon their deaths exploded, flinging the heavy elements as dust into the cosmos. Out of this stardust of light and heavy elements other stars and planets have formed and continue to form today. One of the stars that formed 9 billion years after the big bang was our Sun, with several orbiting planets.

On our planet, Earth, the environment for the first billion years was harsh and chaotic as elements—both light (hydrogen and lithium) and heavy (oxygen, nitrogen, phosphorus, iron, and others)—underwent many cycles of chemical mixing, heating, cooling, and bonding. Chemical molecules by the billions formed, mixed, disassociated, and mixed again, and from this dynamic stew, a community of common ancestral molecules arose, including nucleic acids: strands of ribonucleic acid (RNA) and deoxyribonucleic acid (DNA). A metabolism mechanism formed that could capture energy from the environment and with it the basic molecular building blocks of the first self-replicating molecule evolved and life appeared on Earth. The biologist J. Craig Venter sees the DNA in the cells of animals and humans as a history of the years of

biological evolution by a self-replicating bag of chemicals.

It is not certain whether the molecules of life formed in little warm ponds as envisioned by Charles Darwin or in another environment, such as hot underwater volcano outlets. However, from a bare beginning, molecular life evolved, struggled, and survived. Three life forms evolved: archaea, bacteria, and eukarya, this last the branch of the tree of life to which we belong.

Biological evolution is now understood to be more complicated than Darwin's vision of a single tree of life with all life on Earth evolving from one common ancestor. This is due to the discovery[39] in 1977 by microbiologist Carl Woese that life consists of not just one or two primary domains or branches of the tree, but three. Woese also discovered that bits of mutated DNA can be inherited by horizontal gene transfer across species as well as by the known Darwinian vertical gene transfer from inheritance. This means that a more accurate tree of life would have to include branches that fused, but we will avoid this detail and stick with Darwin's simple version of the tree of life, the eukarya branch.

Since we are focused on the evolution of *Homo sapiens*, a vertebrate, the eukarya domain, our branch of life, will be the center of our discussions. From this main branch, many smaller branches have sprouted, and many others have been pruned over time by the Darwinian requirement of survival. Our branch, that of the vertebrates, survived and

---

39. David Quammen, *The Scientist Who Scrambled Darwin's Tree of Life*, New York Times, 2018.

grew with the evolution of many new species, which after 3.8 billion years includes *Homo sapiens*.

Understanding human evolution requires that we understand it at the micro level of the DNA and RNA molecules. The human genome is comprised of more than three billion base pairs,[40] so studying DNA requires handling a large amount of data. The basic chemical assembly of the DNA and RNA molecules obeys mathematical bonding theories of chemistry.

Technology has given scientists the ability to obtain data at this basic molecular level. New lower-cost DNA-sequencing machines provide rapid computer analyses that have dramatically accelerated biological research, allowing scientists to study the DNA of not only Homo sapiens but of other species as well. The analysis of the vast amount of DNA data generated from the study of the simplest bacteria to the entire genome of animals as well as humans is possible only with the use of supercomputers. These tools are giving scientists a base for understanding the functioning of individual genes and indeed life itself at the molecular level.

The increased capability to sequence DNA cheaply has enabled biologist to study details of evolution at the molecular and gene level that Darwin could only have dreamed about 150 years ago. Which gene gives which trait—long legs, height, skin color, etc.—are questions being answered in lab experiments.

The DNA from the Neanderthal fossil discovered during Darwin's time has been sequenced and the DNA differences compared to those of *Homo sapiens*. He would have been thrilled to learn that the genome of our close ancestor, the chimpanzee, is 98.6% the size of the human genome; we are not quite kissing cousins. But the Neanderthal genome is 99.7 percent the same as that of *Homo sapiens*, making them pretty close to a kissing cousin.

Experiments have been conducted to study how the building blocks of life, amino acids, could have been formed from "the bag of chemicals" on the new planet Earth. One early experiment[41] with the initial (postulated) conditions of Earth 4.5 billion years ago, a version of Darwin's little pond,

---

40. A base pair is a unit consisting of two nucleobases bound to each other by hydrogen bonds.
41. Stanley L. Miller and Harold C. Urey, *Organic Compound Synthesis on the Primitive Earth*, Science, no. 3370, 1959.

was simulated in a laboratory test.[42] After a short time, the lab experiment produced some, but not all, of the amino acids that would be needed for RNA and DNA molecules; that is, for life. This was only a small first step, but it opened the door to further experiments on how life evolved from elementary chemical elements to amino acids and to more complex double-helix DNA molecules that could self-replicate.

From the first self-replicating molecule that evolved, the evolutionary process described by Darwin's theory of natural selection (with modifications) explains the growth and diversification of biological life on Earth that has evolved over the last 3.8 billion years and continues to evolve today. Darwin's theory had proven to be a biological law of Nature.

Man has attempted to understand the details of his biology for thousands of years—how our body really functions. But early on, man had no tools that could be used to study life other than his own eyes, which could observe changes from disease or injury. Medical problems were being addressed without understanding the details of the processes of life. Slowly, as the tools for research became available, an understanding of these processes and the basic units of biological life emerged: the cell DNA (the record of life in each cell), the metabolic process (the supply of energy for our cells), the gene (the unit of heredity), and the chromosome (a collection of many genes). The body also uses larger biological systems that are made from clusters of many cells that operate as a unit to accomplish complicated functions necessary for life, such as the brain, the eye, the heart (the pump for blood circulation), the clotting system for repairing leaks in the blood circulating system, and so on. These and other organs support a brain that monitors the body's functions, reacts to external stimuli, and becomes aware of and reacts to our environment. Many biological systems are necessary for a complicated life form to function.

The evolution of life on Earth has been lucky, for during its 3.8 billion years Earth's environment has been shocked by several sudden mass extinction events that have killed or otherwise significantly affected many species. So, it is by luck that our small furry ancestors were in the right niches on Earth at the right time to survive during mass extinctions that have hit the Earth.

---

42. In 1871, Darwin described a hypothetical warm little pond rich in chemicals and salts with sources of light and heat (the Sun) and electricity (thunderstorms). He imagined that in such an environment, proteins might spontaneously form and become the basis of something more complex and form the basis for evolution.

For the last several thousand years, the information gathered to understand biological life has been influenced by the religious and philosophical beliefs of churches and rulers, which at times hindered the advance of knowledge. However, a major step was taken in the 1670s when a scientist, a Dutchman named Anton van Leeuwenhoek, was able to see life at the microbe level when he constructed the first microscope that allowed life to be seen at the cellular level. He was able to show that life was more complicated than as described by church dogma.

Many other biological discoveries have followed, including the existence of discrete inheritable units of life (genes). A gene is a sequence of DNA or RNA molecules that codes for a molecule that has a function. This major function was discovered experimentally in the 1880s by Gregor Mendel, who was studying inheritance patterns in common edible pea plants by tracking distinct traits from parent to offspring. He found that each discrete trait (height, seed color, etc.) might turn up in combination with any other, for each was transmitted independently. In 1905, the traits Mendel had observed were identified as genes by the Danish botanist Wilhelm Johannsen.

Another fundamental discovery, cell replication by the double-helix DNA molecule, was made by Francis Crick and James Watson in England in the 1950s. DNA was shown to be the molecular repository of the genetic information of life. Insight into the structure of the double-helix DNA molecule was provided by Rosalind Franklin and Maurice Wilkins using X-ray crystallography, and from these data Crick and Watson discovered the double-helix structure: a double-stranded DNA molecule whose paired nucleotide bases provided the mechanism of genetic replication. The details of the functioning of DNA are far more complicated; biologists now know that during gene expression, the DNA is first copied into RNA, which can be directly functional or an intermediate template for a protein that performs a function.

As mentioned, in 1859, Charles Darwin provided the theory of the evolution of life by natural selection with the publication of his book *On the Origin of Species*, which fundamentally altered man's vison of biological life. His theory explained the mechanism that underpins the evolutionary biological change of one species into another: adaptive variations that are helpful to the survival of life in changing environments lead to new species. This is the process by which primitive forms of life evolve to more complicated forms. Details of these changes have now been identified as changes or mutations of the DNA molecule within cell genes. Darwin's theory (with modifications) explains the wide diversity of life forms we see today on Earth created over the 3.8 billion years of the evolution of life from the first self-replicating molecule.

Over the last fifty years, refinements to Darwin's theory have been made, one of which includes the social impacts of groups on survival. A leader in this modification to the theory of evolution is E. O. Wilson,[43] who notes,

> *The effect on the survival and reproduction of the individual is called inclusive fitness. In the second, more recently argued theory is multi-level selection. This formulation recognizes two levels at which natural selection operates: individual selection based on competition and cooperation among members of the same group, and group selection, which arises from competition and cooperation among groups.*

Inherent in this is the importance of social interactions, also noted by Wilson:

> *Natural selection for social interaction—the inherited propensities to communicate, recognize, evaluate, bond, cooperate, compete, and from all these the deep warm pleasure of belonging to your own special group. Social intelligence enhanced by the group selection made Homo sapiens the first fully-dominant species in Earth's history.*

From the study of group selection, we get an insight into the lingering strength of tribalism in animal and human affairs. Grouping by religion (god) or by government (state) to the exclusion of others is an outgrowth of exclusive social expressions also practiced by tribes. Sociability among vertebrates has tempered the tribal urge for exclusion and provided a base for a common morality of all peoples.

Life's evolutionary steps have also been observed in physical changes in fossils, mostly from the bony parts that are more likely to survive. Fossils have been discovered that offer highlights of evolution from worms to fish, amphibians, reptiles, birds, mammals, and us, *Homo sapiens*. Human evolution is a tested fact: Darwin's theory tells us how it works, fossils show us how it looks, and DNA gives us the details of how it works.

---

43. Edward O. Wilson, *On Human Nature,* Cambridge, MA, Harvard University Press, 1978.

Darwin's theory is a problem for those proposing religious theories to explain that man (that is, his creation) is not a product of Nature through the process of natural evolution but rather of supernatural creation by a god as discussed in religious narratives. Most scientists, including Alfred Russel Wallace, who accomplished extensive fieldwork in evolution during Darwin's time, have come to support Darwin's theory.

Biological evolution has produced a broad, bushy tree of life with great diversity. The evolutionary branch of the tree of life of interest here is that of the vertebrates, which has a long history—from worms to hominins to *Homo sapiens*. The growth of the brain and its increased functionality is a notable feature of hominin evolution and a major contributor to the survival of species, which occurs not only from breeding, toolmaking, and hunting, but also from the acquisition of social skills that foster individuals' ability to survive as members of tribes. Brains increased in size during hominin evolution as they have grown in functionality, becoming capable of self-awareness, symbolic thinking, sociability, planning, and language. Other animals have evolved some of these characteristics, but the evolution of hominins has expanded them to such a level that *Homo sapiens* have become the dominant animal on all continents.

Central to hominin brain growth has been an evolved mental capacity to be aware of the surrounding world and able to engage in symbolic thinking and conceive of supportive social groupings. Studies of social interaction among animals and hominins are providing insights into the evolution of individual and group morals that underpin the emergence of large social groups. This functional capability of the brain has given *Homo sapiens* the mental ability to expand its knowledge of Nature in many fields of science (among them astronomy, physics, and biology). This knowledge has provided a base understanding of the evolution of life on Earth.

Unfortunately, many people are still grappling with the science of evolution. Those who hold that a supernatural god created everything have a particularly hard time accepting the facts that have been discovered about Nature's creations over the last 4 billion years.

Early Life

Earth's atmosphere contained little oxygen when it formed 4.5 billion years ago. It was in this environment that the first self-replicating molecule appeared after 700 million years of evolution. As mentioned previously, research has led to the discovery of not just one, but three branches of life (archaea, eukarya, and bacteria) that evolved. We do not know which was first. A summary of the sequence of early evolution in the eukarya branch, the one that contains vertebrates, is illustrated in Figure 7.

About 3.8 billion years ago, single cells evolved, and 2.8 billion years ago tiny organisms with multiple cells, known as cyanobacteria or blue-green algae, evolved. Over the next billion years they proliferated and through their photosynthesis, chemistry—using sunshine, water, and carbon dioxide—produced carbohydrates and oxygen, changing the atmosphere on Earth into an oxygen-rich (21 percent) environment.[44] This environment supported the evolution of aerobic organisms that began to consume oxygen, bringing about an equilibrium in oxygen availability on Earth. Around 1.9 billion years ago, eukaryotes (cells with a nucleus) formed, and more complicated multi-celled organisms formed a little later, about 1.6 billion years ago. Around 1.2 billion years ago, sexual reproduction evolved and provided a biological function that accelerated the evolution of life. Around 700 million years ago, rapid diversification of complex organisms gave Earth many new and different life forms, including the vertebrates. The time of this diversifying of life forms is known as the Cambrian explosion.

The great diversity of life may be illustrated by the many long and complex evolutionary paths that have been followed since that time. Consider two widely different animals, humans and octopuses.[45] They have in common large brains and large eyes and both exhibit complex behavior, but their physical differences are great: humans are vertebrates with a bony structure with two arms and two legs and a central nervous system controlled by a brain, while octopuses have no bones but have eight arms and a nervous system consisting of a brain and some brain functions in each arm. Their common ancestor was in the Cambrian period before the split of the animal branch into vertebrates and invertebrates, which include octopuses. The eyes are an example of parallel evolution leading to comparable results from different beginnings.

---

44. Lawrence Krauss, *Atom: A Single Oxygen Atom's Odyssey from the Big Bang to Life on Earth . . . and Beyond*, Boston: Little, Brown and Company, 2001.
45. Peter Godfrey-Smith, *Other Minds: The Octopus, the Sea, and the Deep Origins of Consciousness*, New York: Farrar, Straus and Giroux, 2016.

Several of the major animal lineages that have evolved have been preserved as fossils dating from 550 million years ago that have been found at the Burgess Shale fossil site discovered in Canada by Charles Walcott in 1886. There are other similar sites around the world containing many of the same fossils. This period of the diversification of life forms from the Cambrian explosion is unparalleled in the evolution of marine animals.

Vertebrates

Vertebrates first evolved during the Cambrian explosion 500 million years ago. The emerging vertebrate body plan included a notochord, a rudimentary vertebra, a well-defined head, and a tail. This layout has persisted throughout vertebrate evolutionary history.[46]

The years from the first vertebrate to today have been turbulent times, with the Earth undergoing several ice ages and suffering five major extinction events: 450, 375, 252, 201, and 66 million years ago. These events caused worldwide species reductions, up to 90 percent at times, and each greatly altered the evolutionary trajectory of many species.

A major extinction example is the K-T event 66 million years ago. It was caused by a large asteroid (7.5 miles across) hitting Earth, leading to the extinction of 80 percent or more of the plant and animal species, including all large dinosaurs. Putting this meteorite explosion in modern nuclear weapon terms, it was equivalent to a 100-million-megaton bomb explosion.

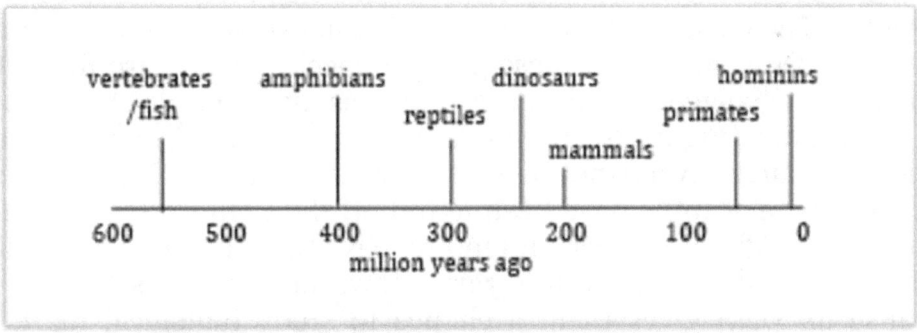

Figure 8 Evolution of Vertebrate Life

---

46. *Vertebrates: Fossil Record,* University of California Museum of Paleontology, February 2005.

Evolution molds life forms by natural selection that favors the animal than adapts to a change and has higher survival odds. This can be caused by mutations in the DNA and lead to a new species. New niches for life appeared in the changed environment after the extinctions, giving some species a better chance for survival and eliminating those with a lower chance. Dinosaurs had been successful and were the dominant land animal on the planet for over 175 million years until the K-T event killed all large dinosaurs. Some small dinosaurs did survive and some of these are the ancestors of today's birds.

A small shrew-like (mouse-size) mammal that evolved 200 million years ago also survived the extinction. But it took a million years for the ecosystem, the plants and trees, to provide enough nourishment for larger animals over 100 pounds to evolve. The rest, as they say, is history as the little vertebrates prospered, diversified, grew in size and evolved into many different species over the next 65 million years, including the primates and hominins (including great apes and *Homo sapiens*).

A 600-million-year section of the vertebrate branch[47] of the tree of life is illustrated in Figure 8, which gives several examples of the diversity of vertebrate life forms over time. The first vertebrates and fish evolved about 550 million years ago during the Cambrian explosion, amphibians 400 million years ago, reptiles 300 million years ago, dinosaurs 240 million years ago, mammals 200 million years ago, primates 55 million years ago, hominins 8 million years ago, and *Homo sapiens* 700,000 years ago.

In the oceans, some life survived the mass extinctions and continued to evolve. The coelacanth, a lobe-finned fish, evolved about 400 million years ago. Descendants of these ancient species of fish are living today relatively unchanged, and they are known as living fossils. An example is the coelacanth (Figure 9). It was first been seen in 1990 after one was accidently caught by a fisherman off the coast of Africa. Since then, small pockets of coelacanths have been found off the Comoro Islands and the east coast of Africa. Fishermen tend to leave them alone, for they live in deep water and are apparently rare. However, biologists love this fish because its DNA gives them a chance to study biological markers of ancient fish that can be compared with the fish of today.

---

47. Walter Holbrook Gaskell, *Origin of Vertebrates,* Sagwan Press, 2015.

Figure 9 Coelacanth—Living Fossil

Vertebrate evolution involves changes in body organs (heart, lungs, etc.) from fish to man over hundreds of millions of years. On land and in the sea, new species evolved, and some of their descendants are familiar to us, such as alligators (reptiles), birds (dinosaurs), hominids (gorillas), and hominins (chimpanzees, bonobos, and man). An early example is the evolution of amphibians. About 400 million years ago, some fish evolved into amphibians, making the transition from life in water to life on land. An example amphibian is the fossil discovered by Neil Shubin[48] named *Tiktaalik*, which had a basic fish body type for movement in the water, primitive lungs and gills for oxygen, and primitive fins that allowed limited land movement.

Studies of biological systems (heart, blood clotting, lungs, etc.) have outlined the step-by-step DNA evolution of these biological systems. An example is a critical biological system of vertebrates, blood clotting. The evolution of this system has been studied over the range of the last 500 million years by evolutionary biologists[49] who have illustrated the evolutionary changes from the simple blood-clotting systems of the early hagfish to the more complicated clotting system found in later vertebrates.

On our branch of the tree of life, the evolution of vertebrates included the primates from 65 million years ago. Primates evolved large brains relative to other mammals and lived in groups of up to a hundred members, which required the evolution of substantial social structures. Sociability, cooperation, and brainpower evolved and came together in the primates, suggesting a correlation between the size of primate groups

---

48. Neil Shubin, *Your Inner Fish,* New York: Pantheon Books, 2008.
49. Russell F. Doolittle, *Evolution of Vertebrate Blood Clotting,* University Science Books, 2013.

and the size of their brains. They also differed from their ancestors by having increased reliance on vision at the expense of smell, the dominant sensory system in most mammals.

Following the primates were the hominins, who comprise four genera: humans, chimpanzees and bonobos, gorillas, and orangutans. About 17 million years ago, the common ancestor to gorillas and chimpanzees lived in a forest environment in Africa. About 10 million years ago the climate became drier, and their forest homes were replaced by grasslands, where the chimpanzees split from the gorillas. The bonobos split from the chimpanzees about a million years later.

In this new environment about 7 million years ago, another split occurred, and our ancestor, *Ardipithecus*—chimpanzee-like, walking on its knuckles—evolved. Within several million years, this was followed by the evolution of *Australopithecus*, a bipedal ancestor species along our tree of life.

We look at chimpanzees[50] and bonobos[51] today as "living fossils," with which we can study the evolution of our ancestors. Only about 1.2 percent of the DNA of chimps[52] and bonobos is different from that of humans, but their brain size is one-third that of man. In comparison to humans, chimpanzees are limited in social skills, have poor social-referencing abilities, and rarely engage in general imitation and active teaching.

Young chimpanzees possess exceptional working-memory capacities, superior to those of human adults. However, their ability to learn the meaning of symbols is relatively poor. Chimps have acquired vocabularies of two hundred words or so, and they can even link pairs of words in new patterns. But their vocabularies are small, and they do not use syntax or grammar, the rules that allow us to generate a variety of meanings from a small number of verbal tokens. Their linguistic ability seems never to exceed that of a two-year-old human, and that is not enough to have competed with subsequent ancestors of the *Homo* species.

---

50. Barry Bogin and Carlos Varea, *Evolution of Human Life History*, Jon Kaas, ed., *Evolution of Nervous Systems*, Oxford: Elsevier, 2017.
51. Bonobos split from chimpanzees about five million years ago.
52. Daniel J. Fairbanks, *Relics of Eden: The Powerful Evidence of Evolution in Human DNA*, 2007.

Figure 10 Chimpanzee Family

We have learned that chimps and bonobos are not simple brutes; they are sensitive, intelligent, and social. They have demonstrated a high degree of social intelligence, sense of self, and bonds of care and independence.[53] Their ancestor, the gorilla, had their social rules. Observers of wild gorillas have noted that they are a spirited, tight-knit social group. Dian Fossey, who extensively studied gorillas, said:

> *How many fathers have the same sense of paternity? How many human mothers are more caring? The family structure is unbelievably strong.*

The chimp alpha male becomes the leader of the tribe, and in doing so he forms a small social group that aids him in his necessary job of keeping peace within the tribe, a necessary state for survival. Although fights for dominance occur, they are usually contained by social means. But chimps are known to kill other chimps on occasion. With bonobos it is the females who are dominant in their tribes, which exhibit less aggression. They exhibit more peaceful tribes and are not known to kill other bonobos.

Chimpanzees and bonobos exhibit strong social bonds among their family and group; in effect, their primitive social rules are an early form of social morality. Frans de Waal notes:

> *The only possibility (to explain their social rules) is to embrace morality as natural to our species.*

---

53. Jeff Sebo, *Should Chimpanzees Be Considered 'Persons?'* New York Times, April 7, 2018.

Frans de Waal has also observed that sense of social caring in other animals:

> *I've argued that many of what philosophers call sentiments can be seen in other species. In chimpanzees and other animals, you see examples of sympathy, empathy, reciprocity, a willingness to follow social rules. Dogs are a good example of a species that have and obey social rules; that is why we like them so much, even though they are large carnivores.*

Thus, it can be concluded that morality has been an evolving, socially driven characteristic of vertebrates that *Homo sapiens* inherited and expanded.

Chimps and bonobos, with their social characteristics, have demonstrated a degree of generosity and trust that has not only assisted in their survival but also given a base for human generosity.[54] Figure 10 is a photo of a chimpanzee family that was part of an experiment on generosity of food-sharing among these hunter–gatherers.

Fran de Waal reports an encounter of a trainer with an old chimp named Mama[55] that he had studied for a number of years. The trainer heard she was dying and, having not seen her for several years, went to visit her. When he walked up to her, she was obviously very sick, but she broke out in a big smile, put her arm out, hugged him, and stroked his hair. She died the next day. For de Waal it was a very moving encounter.[56]

But there is also a dark side to our early ancestor's social life, for chimps have been observed to band together and at times kill other chimps who have become socially unattached to the main group. Researchers at the Jane Goodall Institute have reported a group of chimpanzees attacking and killing a smaller group that had left the main tribe. It appears that hominin tribalism has had an early and deep implant in *Homo sapiens* ancestry.

The evolutionary step of chimps to *Homos* appears to have been a drawn-out process with many encounters between several evolving species before *Homos* arrived. For simplicity, the evolutionary links from chimps to *Homos* are summarized by two genera, *Ardipithecus* and *Australopithecus*, with the note that evolution is not a simple one-to-one

---

54. Carl Zimmer, *Seeking Human Generosity's Origins in an Ape's Gift to Another Ape*, New York Times, September 11, 2018.
55. Frans de Waal, *Mama's Last Smile*, New York: W. W. Norton and Company, 2019.
56. This encounter can be seen in a YouTube video, *Mama's Last Hug*.

link of evolving species but rather a many-interrelated-species entangled mess. The following are representative examples, and some evolving species have been left out.

Ardipithecus

Ardipithecus evolved about 6.0 to 4.5 million years ago and lived among their chimp neighbors. Their fossils suggest that they retained many chimp-like features; they were hairy with small brains (300–350 cm^3, one-third those of *Homo sapiens*), but they also showed the beginning of the transition to more humanlike features. Their bone structure had some humanlike toe and pelvic structure and an apparent cranial expansion of vocal capabilities, an early step toward more complex verbal communications.

The fossil Ardi, a female *Ardipithecus*, dates from 4.5 million years ago and suggests when compared to other fossils that females and males were almost equal in size. Although she was a biped, Ardi's toes and thumbs enabled her to also climb trees easily. Further, her stature, smaller teeth, and less projecting face suggest less reliance on aggressiveness and more on socialization within the tribe for survival. This trait of socialization on the vertebrate branch appears to be important to vertebrate evolution and extends to gorillas, chimps, bonobos, and humans.

Australopithecus

The species *Australopithecus*[57] evolved about 4.5 million years ago and was around for several million years, until about 2 million years ago. During this time, they were adventuresome, with some journeying of tribes as far as Europe and Asia. Their skeleton structure suggests an evolutionary transition from an apelike to a humanlike walking gait. Their fingers, however, were still curved like those of tree climbers, but they were capable of crafting tools: at one site dating from 3.3 million years ago, rudimentary stone tools were found along with the fossil of a young *Australopithecus* female.

Three example (Figure 11) *Australopithecus* fossils—labeled "Couple," "Selam," and "Lucy"—give insights into the species. The Couple, at 3.7 million years old the oldest of these fossils, are footprints of two *Australopithecus* walking side by side with a humanlike gait. This

---

57. *Australopithecus* fossils include *A. afarensis*, *A. africanus*, *A. garhi*, *A. aethiopicus*, *A. robustus*, and *A. boisei*.

footprint fossil was discovered by Mary Leakey at the Laetoli site in Tanzania, a short distance from her earlier fossil finds of *Homo habilis* at

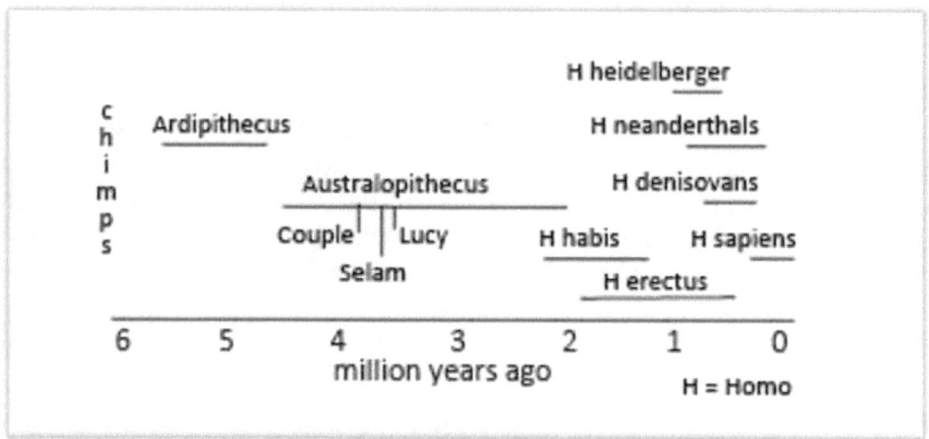

Figure 11 Evolution of Hominins

the Olduvai site. Figure 12 is a photo of a cast of an eight-foot section of the seventy-five-foot-long fossil showing a set of parallel footprints of two *Australopithecus*, one made by a larger and heavier individual, assumed to have been male, and the other by a smaller and lighter companion, assumed to have been female. The footprint fossil of the two walking peacefully gives support to a transitional date of about 4 million years ago for the evolution of the human walking gait from the chimp-like lumbering gait to humanlike bipedalism.

This footprint fossil is personally notable to me. My interest in evolution led me to follow my wife to Africa. The head of the San Diego Zoo's Wildlife Preservation Department had plans to take a group on a safari to Kenya, and she wanted to sign us up. The problem was me, a city boy who thought there were other trips more appealing than bouncing around in off-road vehicles and sleeping in tents on the Serengeti with lions wandering about. But she insisted, so I invented a ploy to get out of going; I said I would go on the condition that the trip included having tea with Mary Leakey, a famous scientist who at that time was at the museum in Nairobi where several of her discoveries were exhibited. Leakey was famous for her fossil discoveries at Olduvai Gorge and at Laetoli. I was certain she would be far too busy and, in any event, uninterested in having tea with a small group of duffers from San Diego.

Figure 12 *Australopithecus* Fossil Footprints—The Couple

A week later, my wife got a telephone call from the zoo's trip leader announcing that Mary Leakey had agreed, and the trip was on. The leader was an old friend of hers and had worked with her in the past, so he had simply telephoned, and they set a date for our visit. My ploy having spectacularly failed, in a few weeks we were airborne, heading to London and then on to Kenya.

Our meeting with Mary Leakey was set to take place at the Nairobi Museum several days after we landed in Kenya. Our little group of eight arrived at the museum on time, but she was not there. The museum curator serving as our guide said that we would all soon know when she had arrived, and indeed, we saw and heard people in the parking lot running for their lives when she drove in with reckless abandon. Her reputation of being the worst driver in Kenya was intact.

We met her at the fossil displays, which included her and her husband's (Louis Leakey's) discoveries. Tea was served, and she proved to be a gracious hostess to the San Diego duffers. On display was a copy of a skull fossil (*Homo habilis*) and other fossils she had found at Olduvai Gorge dated back 1.8 million years and an eight-foot plaster copy of a section of the fossil of *Australopithecus* footprints that dated back 3.7 million years, which she had discovered at Laetoli.

She talked about the social evolution of the *Australopithecus* occurring 4 million years ago with bipedal walking and the use of stone tools about halfway along the journey from our chimp ancestors to *Homo sapiens*. Her view was that the footprints were made by two *Australopithecines* walking side by side, upright, one being larger and heavier than the other—in essence, a couple. The stride of the two suggested that they walked with a humanlike gait and were about four-and-a-half feet tall. After fifty feet the footprints show that they paused and made a left turn and stopped. They resumed walking in the original direction at the same pace. Whatever they saw during their pause did not panic them, for they resumed their walk at the same steady pace. Possibly, they saw a small animal or a volcano plume appearing in the distance, or perhaps they were just looking for food. Their footprints were preserved soon afterward by a deep layer of falling volcanic ash.

Mary Leakey thought that these two small hairy *Australopithecus* leaving those footprints were our ancestors peacefully doing their daily chores and that they were an important link in our evolution. How could we not have a feeling of togetherness with them? It was just one little visible step on the journey that would take many generations over the next 3 million years for the evolution of *Homo sapiens*. These fossil footprints suggest that an important evolutionary step toward humanization had been made by this small hominin with a brain size of 400–550 cm^3, a third that of *Homo sapiens*. The ability to walk bipedally freed the hands to accomplish more difficult tasks, such as making tools, and improved running speeds. These were capabilities beyond those of the chimps.

The second *Australopithecus* fossil example (Figure 13) is the skull of a three-year-old female called Selam, which has been dated to 3.3 million years ago. At the fossil site, crude stone tools[58] were also found. The skeleton supports the structure required for a humanlike gait. The fossil shows the link to ape ancestors, with shoulder blades, arms, and fingers for climbing like those of a chimpanzee, but its teeth and skull are more humanlike.

---

58. John Noble Wilford, *Stone Tools from Kenya Are Oldest Yet Discovered*, New York Times, May 20, 2015.

Figure 13 *Australopithecus* Fossil—Selam

The third fossil noted is Lucy, which has been dated to 3.2 million years ago. Being an almost complete skeleton, Lucy is the most famous of these three *Australopithecus* fossils. Her bone structure, too, is suited to a human walking gait, but she too retains many chimp features. *Australopithecus* represent our ancestors, little and vulnerable with small brains and some physical traits of their ancestors. Individually, they were weaker than many of their predators, and it would take thousands of subsequent families evolving along our family tree over the following 3 million years to obtain the mental capabilities and social skills of *Homo sapiens*.

Homos

The earliest of the *Homo* genera were brainy by primate standards, although their brains were in size more like chimpanzee or bonobo brains, with a volume of about 300 to 450 cm^3. Our brains today, in comparison, average about 1,350 cm^3. Not only did the brain grow during *Homo* evolution, but some regions of the brain, such as the frontal lobe, grew more than others. This is the region that provides specific functionalities, such as self-awareness, symbolic thinking, and planning. Other animals had evolved some of these capabilities with limited functionality, but the evolution of *Homos* expanded these capabilities to such a level that *Homo sapiens* would become the dominant animal.

Our knowledge of *Homo* evolution is based on fossil discoveries and, more recently, on DNA analysis of evolution of several fossils where some DNA was recoverable. The task of determining the exact line of evolution is difficult, for over thirty different hominin fossils have been discovered. Our discussions will focus on five *Homo* genera over 500 thousand years that illustrate the range of species evolution leading to the *Homo sapiens*.

Seven *Homo* species are discussed[59] as representative examples of the complicated *Homo* evolution: *habilis, erectus, heidelbergensis, neanderthalensis, denisova, floresiensis,* and *sapiens*.

A summary timeline for the evolution of these example *Homos* is given in Figure 11. These examples reflect the fossils discovered, and as new fossils are found this timeline and the order of the ancestral links may be modified. Although the figure gives the impression of a tidy sequence of evolution from chimps to man, the breadth of the data from the fossil discoveries suggests that the *Homo* branch is more like a bush, with many closely related and interrelated species complicated by interbreeding between species over several hundred thousand years. The fossil discoveries give the features of physical changes, such as the skeleton, brain size, and so on, but the small number of fossils limits our ability to find a species average and the range of variability necessary to determine the exact line of evolution.

The evolution of the brain's capabilities is also complicated, but an evolutionary analogy may be helpful. Think of an ancestor with limited sight who had the capability to see only fuzzy black and white images. His children could do a little better—they could see the black and white images sharper. Their children did even better by seeing some of the colors that made up the white images, say, red and blue. Following generations could see even more hues. At each step of this progression, the expanded capabilities of their improved sight gave them an evolutionary advantage over others whose sight was more primitive. The same is true of the evolution of the brain's capabilities from, say, our ancestor the gorilla to us today; we who can perceive more clearly the vast world around us and the complicated social environment in which we live. The additional capabilities of *Homo sapiens* required larger brains, and that is what has been found in the *Homo* fossils along our branch of the evolutionary tree.

The changes in stone tools found at fossil sites provide insights into the growth of *Homos*' mental capabilities. As *Homo* tribes grew larger, they also evolved an ability to make more sophisticated stone tools. And they developed more sophisticated social concepts,[60] as suggested by the growth of the brain's frontal lobe. The brain size doubled from 600 cm^3 for *Homo habilis* to 1,250 cm^3 for *Homo sapiens*.

---

59. Other *Homo* species have been omitted, such as *Homo naledi* and others.
60. Rodolfo Llinás and Patricia S. Churchland, eds., *The Mind–Brain Continuum*, Cambridge, Massachusetts Institute of Technology, 1996.

Homo habilis

The *Homo habilis* genus followed *Australopithecus*. The first *Homo habilis* evolved around 2.5 million years ago and disappeared about 1.5 million years ago. Their fossil sites suggest that they[61] had developed more efficient stone tools with sharper edges for hunting and food preparation and had a broader diet than the earlier *Australopithecus*. *Homo habilis* began to become significantly more intelligent, with a larger brain (550 to 650 cm^3) than that of *Australopithecus*.

A cognitive evolutionary schedule developed by Fuller Torrey[62] can be used as an outline of the *Homo* species brain evolution. *Homo habilis* had a brain size 50 percent larger than that of *Australopithecus*, with a marked increase in the growth of the cerebral cortex. This difference shows an increased mental capability to make tools of greater complexity, even to make one tool to make another tool. The cerebral cortex region is considered to be the center of executive functions, including planning complex behavior and controlling social behavior. Further, it is thought that *Homo habilis* was probably the first to have developed a primitive proto language. These are measures of intelligence and the enlargement of this region of the brain would continue in subsequent *Homos*.

Homo erectus

Homo erectus appeared about 2 million years ago and was, in body, one of our first truly humanlike ancestors, and they were the first to spread out to Europe and Asia. They were taller than *Homo habilis*, some of them as tall as modern humans, with a brain three-fourths the size of man's. They made more sophisticated stone tools than *Homo habilis*, including beautiful, carefully designed stone tools known as Acheulean axes. Better stone tools may have given *Homo erectus* access to more meat, a crucial source of high-energy food for their larger brains. They may have also learned to manage, control, and use fire, which would have allowed them to tap into a huge new source of energy, cooked food. However, the brain functions of *Homo erectus* did not progress to further inventiveness. Over the 2 million years of evolution of the *Homo erectus* species, the stone tools remained about the same.

---

61. Several discoveries of hominin fossils suggest that these may have been earlier hominins, but more finds are needed to substantiate this.
62. E. Fuller Torrey, *Evolving Brains, Emerging Gods: Early Humans and the Origins of Religion*, Columbia University Press, 2017.

In addition to being more intelligent, it is likely that *Homo erectus* were also self-aware and aware of what others were thinking about them. Self-adornment can be a means of advertising one's family relationships, social class, group allegiance, or sexual availability. The evolution of an introspective self would have provided major cognitive advantages over other *Homos*, especially in social interactions and in being able to predict others' behavior and facilitate cooperative living in larger groups. This would have greatly increased the efficiency of group endeavors, such as group hunting, and put *Homo erectus* at a significant advantage in warfare against others who did not possess this degree of cognitive skills.

Homo erectus prospered in Africa and were the first *Homo* to walk out of Africa; this occurred around 1.7 million years ago, as shown by their fossils found in distant places, such as Georgia (Crimea) and the Far East (China and Indonesia). The venturing of *Homo erectus* out of Africa probably could not have happened without the elevated level of social cooperation and communications with a proto language. They were around more than a million-and-a-half years, disappearing about 120,000 years ago.

A pocket of small (three-and-a-half-feet-tall) *Homos* was discovered on an island in Indonesia; they appear to have been an ancestor of *Homo erectus*. They have been given the name *Homo floresiensis* and lived on the Indonesian island until about 50,000 years ago. Although they had a small brain, they used fire and made stone tools similarly to *Homo erectus*. Their very small stature, small brain, and distant location make them the "out of place" and "out of time" *Homos* who remain poorly understood.

Homo heidelbergensis

An ancestor of *Homo erectus* was *Homo heidelbergensis*, who evolved in Africa about 800,000 years ago and lived until 300,000 years ago. They also migrated out of Africa, and their fossils had been found in southern Africa, east Africa, and Europe.

The brain size of *Homo heidelbergensis* was about that of *Homo sapiens*. It is argued that *Homo heidelbergensis* used broader language than its ancestors. The evolution of improved speech would have improved tribal communications and thus increased socialization, allowing for larger tribes and improved survivability over contemporary *Homos*. The stone tools found with the *Homo heidelbergensis* fossils are sophisticated and suggest that they could make fires, craft complex stone tools, and fashion wooden-handled, stone-tipped hunting spears. Robin

Dunbar has argued[63] that the critical event in the evolution of the neocortex took place with *Homo heidelbergensis*, whose neocortex is large enough to develop a more complex proto language and larger social groupings.

Homo neanderthalensis

Charles Darwin's book *On the Origin of Species* was published in 1859, a time when the long-held prevailing view—species were immutable and originally created by a supernatural God—was beginning to be questioned by geologic and fossil discoveries that indicated Earth was much older than dates referred to in the biblical narrative. The Neanderthal fossil discoveries were humanlike fossils that raised pressing questions: were they from a species older than humans, could they be our ancestors, and could Darwin's theory of evolution be correct?

The reaction to the discovery of Neanderthal fossils was an example of the struggle that has occurred from the first appearance of humanlike, but not exactly human, fossils. In Germany in 1856, a fossil skull was found in a quarry that, with its heavy brow ridge, was clearly not from humans, but rather from another species. The second Neanderthal skull was found in 1868, and since then others had been found in other places in Europe. Fossils found in French caves in 1908–11 include almost complete skeletons.[64] The fossil data supported the view that there was a species other than humans, possibly older than humans, and that Darwin's theory could be right. Further research has found that Neanderthals, in addition to *Homo sapiens*, also left a fossil record of tools, jewelry, and grave goods.

Darwin's theory has been shown to be right from fossil discoveries that indicate that Neanderthals split from *Homo heidelbergensis* about 500,000 years ago, Denisovans split about 450,000 years ago, and *Homo sapiens* split about 400,000 years ago. At different times they journeyed out of Africa and into Europe and Central Asia, first the Neanderthals and Denisovans and then *Homo sapiens*.

In parallel with the discoveries of Neanderthal fossils in Europe, discoveries were being made of humanlike fossils that were different from those of humans and Neanderthals in places as different as Java and Africa. Now over four hundred Neanderthal fossils have been identified

---

63. The study is based on a neocortex size plotted against several social behaviors of living and extinct hominids.
64. Ian Tattersall, *The Fossil Trail*, Oxford University Press, 1995.

and dated using geographic and radioisotope dating techniques. An entire genome of a 50,000-year-old Neanderthal fossil bone has been sequenced, giving DNA data for comparisons with human DNA. The data revealed that Neanderthals shared 99.7 percent of their DNA with *Homo sapiens* and hence were much more closely related to us than their closest nonhuman relative, the chimpanzee (98.8%). Complicating the evolutionary picture is that Neanderthals sometimes lived in the same place at the same time as *Homo sapiens* and interbred with them. DNA comparisons clearly show the small differences.

Homo heidelbergensis, Neanderthals, and Denisovans have vanished from the fossil record for reasons not known and it appears that by 40,000 years ago, *Homo sapiens* was the only *Homo* species in the world.

Recent comparison of the DNA and genes of Neanderthals and *Homo sapiens* shows that each has about 20,000 genes and they differ only in sixty-one genes, and among those are only four genes that determine the growth of the brain. If so the four different genes in our brain development made a very large difference in our social behavior.

The last fossil sites of Neanderthals were at Spain's water edge at Gibraltar. Why they died off and only *Homo sapiens* survived is still an open question. One view is that the superior social and cognitive skill made *Homo sapiens* the winner.

Homo Denisovans

Denisovans shared a common origin with Neanderthals, splitting from *Homo heidelbergensis* about 450,000 years ago. They lived until about 50,000 years ago and ranged from Siberia to Southeast Asia. During this time, they lived among and interbred with the ancestors of modern humans. The DNA of aboriginal Australians is about 3 percent to 5 percent DNA from Neanderthals, which, combined with Denisovan DNA, represents 17 percent of the genome.

Homo sapiens

Homo sapiens evolved from *Homo heidelbergensis* about 400,000 years ago in Africa. After that, waves of *Homo sapiens* at 170,000 and 70,000 years ago migrated north to Europe, and in some places they overlapped in time and place and interbred with other *Homos*, Neanderthals and Denisovans. Some of the interbreeding can now be seen in the percentage (1 percent to 4 percent) of Neanderthal DNA found in the DNA of *Homo sapiens* that lived in regions north of the Sahara Desert. Further, in some other places in the world, the interbreeding was greater.

About 200,000 years ago, Fuller Torrey argued, *Homo sapiens* evolved an awareness of what others were thinking, a major cognitive skill. What people think about other people's thoughts and what people think that others think about their thoughts are the core of most complex social interactions. About 100,000 years ago, *Homo sapiens* evolved additional cognitive skills, including the introspective ability to think about themselves thinking about themselves. This mental growth of the *Homo sapiens* brain is observed in the fossils, which show an enlarged cerebral cortex region, and in the fossil record of tools and weapons, self-ornamentation, burial goods, and art and artifacts. Around 50,000 years ago, Steven Mithen[65] noted, it was the speed at which tools and art were modified and new ones introduced that set *Homo sapiens* on a new and rapid evolutionary path that continues today.

The evolutionary changes leading to the emergence of *Homo sapiens* required both brain and body change, and this growth required a large investment of energy and time by older members of the family and social groups in infants and children. This was achieved via a new type of approach to the postnatal life of *Homo sapiens*—adding two stages to a child's growth, a childhood stage after infancy and an adolescence stage after the juvenile stage.[66] The results enhanced reproductive and growth success and increased learning time for the brain development.

Although having roughly the same brain size as their ancestors—*Homo heidelbergensis*, Neanderthals, and Denisovans—*Homo sapiens*' brains evolved differently internally, with a large increase in the prefrontal cortex region of the brain, which suggests increased sociability and mental capabilities, including symbolic thinking, planning, and decision-making. It is thought that *Homo sapiens*' brains acquired the ability to form an understanding of the minds of others and shared intentionality. These capabilities also show the arrival of a modern form of language that better served social functions with more specific communication. The combination of increased social skills, new mental skills, and speech contributed to larger and more informed *Homo sapiens* tribes as compared with smaller tribes of Neanderthals. This gave them an edge in basic hunting and gathering. No animal can swap stories about the future or the past, warn about the lion pride ten miles to the north, or talk about gods or demons. Our species seems to have been the first to cross the

---

65. Steven Mithen, *The Prehistory of the Mind,* London: Thames and Hudson Ltd., 1996.
66. Barry Bogin and Carlos Varea, *Evolution of Human Life History,* in Jon Kaas, ed., *Evolution of Nervous Systems* 2e. vol. 4, Oxford: Elsevier, 2017.

linguistic threshold beyond which information can accumulate within communities and across generations.

The increased cortex size was thought to be linked to an increased ability to plan and socialize, which required the use of long-term memory to anticipate gain down the line. For many years, this was thought to be a uniquely human trait that develops in young children. But it turned out that chimpanzees have this ability, and they also make tools. But their abilities are limited. For reference, four-year-old humans can plan for long-term gain as well as adult chimpanzees can. The little *Australopithecines* couple who left their footprints at Laetoli on the African plains may have had small brains, but socially they had made an important evolutionary step by learning to work better collectively. This was a start and over generations, *Homo* communities hunted and gathered with growing skill and efficiency.

Early evidence of ritual, symbolic, or artistic activity is particularly significant because it suggests the level of ability needed to think symbolically or tell stories about imaginary beings. Fuller Torrey argues that the increase in *Homo sapiens*' mental capabilities continued with subsequent major changes, one about 50,000 years ago and a second about 12,000 years ago. About 50,000 years ago, modern *Homo sapiens* developed an autobiographical memory: the ability to project themselves backward and forward in time, using their experiences from the past to plan the future. This cognitive capability allowed complicated tales to be invented, a necessary ingredient for the religions to come.

Also at this time, improvements were made in the development of tools and weapons. It was in this period that *Homo* social growth became semi-autocatalytic, for brain evolution gave more efficient hunting tools, which gave more food, which allowed more time for increased socialization, which gave more time for teaching children, which increased their mental capabilities to be used to increase the efficiency of hunting and hunting tools and methods.

Further improvement in mental capacity about 50,000 years ago led to a gradual abandonment of the old stone tool technology and replacement of it with wood and bone tools. The bow and arrow, which had been used occasionally in earlier millennia, came into widespread use by at least 20,000 years ago. It was most helpful by allowing a hunter to hunt at a safer distance from prey. Another tool that came into use at this time was the lamp. Controlled use of fire had been possible for over 100,000 years, but the use of lamps was new. Lamps made possible the extensive cave art displays that have been discovered in deep caves dating back to 40,000 years ago. Also around this time, the first musical instrument, the flute, appeared. Several have been found in Germany.

Evolution[67] is adaptive and changes what is there. Not all social features are changed at once. Some of the old chimp-like traits of impulsiveness, hedonism, and inhibition remain in our brains. Some genes in our own genome have been found to be related to Neanderthals, as well as the Denisovans, so *Homo sapiens* has both some of the "old" brain and some of the evolved "new" social brain of these ancestors.

Homo sapiens had successfully competed in the race for survival with other hominins, as well as with a range of formidable animal predators they would have encountered in Africa, Europe, Asia, and Australia and Indonesia (which they reached about 45,000 years ago). The last *Homo* competitor in Europe, the Neanderthals, disappeared from their sites in Europe about 30,000 years ago. In another place in the world, a *Homo floresiensis* fossil was found that dated to 50,000 years ago. None have been found after that date.

---

67. Jesse Bering, *The Belief Instinct,* New York: W. W. Norton & Company, Inc., 2011.

VI. Mental Constructs

Once upon a time on Earth, there were no gods, demons, or angels to talk about, for our early ancestors' brains had not evolved the capability to conceive of such things. Previous chapters discussed the creation of the universe, Earth, and biological life on Earth. This chapter summarizes the evolution of the brain of one of the species, *Homo sapiens,* with the mental capabilities to create mental constructs of complex states with kings and social laws as well as narratives with supernatural beings.

To put the evolution of the human brain in perspective, when our vertebrate ancestors first evolved on our trunk (eukarya) of the tree of life 600 million years ago, they were small worms whose brain was a small cluster of nerve cells, a ganglion, attached to one end of a cord (notochord) that would become the backbone. From these few nerve cells in these worms, the brains of vertebrates grew and evolved mental capabilities to react to the environment and, later, complicated social processes. All of these expanded capabilities aided in their survival.

The evolving vertebrate brain did that job very well, for during its 600-million-year journey from this small cluster of brain cells in a worm with extremely limited mental capabilities, it evolved and grew into the brain of *Homo sapiens* with a billion cells with vast mental capabilities. So, at some point during this long evolutionary journey, the concept of a supernatural god was able to be conceived by the brains of *Homo sapiens*.

The concepts of the gods we know are complicated social constructs, so when and how did our ancestors develop the capability to conceive of them? We know that social intelligence had evolutionary roots going back to our ancestral hominids and even before. Frans de Waal notes that all hominins are social animals with evolved social structures in which each member knows its own place and social order is maintained by certain rules of expected behavior that the dominant group leader enforces with punishment. Chimps and bonobos[68] have evolved a sense of fairness, trustworthiness, and other social traits manifested in tribes. These social

---

68. Frans de Waal, *The Bonobo and the Atheist,* W. W. Norton & Company, Inc., 2013.

traits were passed from one generation to following generations that modified them as tribal societies evolved.

Barbara King[69] notes that the traits necessary for the social constructs of gods include high social intelligence, a capacity for symbolic communication, a sense of social norms, realization of "self," and a concept of continuity. The evolution of *Homo* mental capabilities with these attributes has been accomplished by an evolving social intelligence over many preceding species.

The concept of spirits with supernatural powers helped, protected, and comforted individuals. Once *Homo sapiens* had the mental capability of assigning causation to the things happening around them, spirit stories evolved into proto-religions that included rules to service the god (gods). Nicholas Wade[70] has observed,

> *Like most behaviors that are found in societies throughout the world, religion must have been present in the ancestral human population before the dispersal from Africa 50,000 years ago. Although religious rituals usually involve dance and music, they are also very verbal since the sacred truths have to be communicated. So a religion (and its gods), at least in its modern form, cannot pre-date the emergence of language.*

The sum of tribal social rules and religious rules were essentially the morals of that society. The philosopher Patricia Churchland[71] summarizes the evolution of societal morals:

> *Morality seems to me to be a natural phenomenon—constrained by the forces of natural selection, rooted in neurobiology, shaped by local ecology, and modified by cultural developments.*

We can gain insight into the evolution of *Homo sapiens'* social intelligence necessary for the conceptualization of gods from observation of the artifacts discovered at fossil sites that changed over time. Evolution of language was a necessary companion to increased art capabilities. The oldest known ritual burial of modern humans was around 100,000 years

---

69. Barbara King, *How Animals Grieve,* Chicago: The University of Chicago Press, 2013.
70. Nicholas Wade, *A Troublesome Inheritance: Genes, Race and Human History,* New York: Penguin Press, 2014.
71. Patricia S. Churchland, *Braintrust: What Neuroscience Tells Us about Morality,* Princeton, NJ: Princeton University Press, 2011.

ago. Grave goods found in burial sites were placed there to make offerings to the spirits to meet perceived spiritual needs or possibly smooth the deceased's journey to an afterlife. Grave goods—artifacts found in graves—changed over time and reflected increased sophistication, with the introduction of symbolic art, such as jewelry and carvings. Figurines of humans and animal–humans date back to around 40,000 years ago. These artifacts have been found at many *Homo sapiens* graves and at some Neanderthal fossil sites in Europe, suggesting that symbolic thinking and art had become widespread.

Concurrently, decorative art was appearing in caves throughout Europe. Examples are the paintings made around 35,000 years ago in the Chauvet Cave in France. Early cave art at first illustrated hands of the painter, essentially saying, "I am a person, and I am here." The capability to make art expanded from this simple start to complicated and beautiful representative art at later dates in caves, including the painting of horses and other animals they were observing, hunting, and interacting with.

The creation of art with mixed animal–human statues suggests that thoughts of supernatural spirituality had evolved by around 50,000 years ago. An example is the Lion-man, or Löwenmensch figurine, which was found in Germany. Dating from about 40,000 years ago and thus one of the oldest known fossils of symbolic art, it depicts a half-lion, half-man figure, possibly an early god. It was carved out of woolly mammoth ivory using a flint stone knife. Such artifacts were most likely linked to stories about spirits and gods that had been invented by the spiritual leaders and vocally shared among members of the tribe. Both *Homo sapiens* and Neanderthals placed items in graves in homage to the spirits, and over time, simple burials transformed into ritual burials with more elaborate art reflecting the increased sophistication of the stories being conceived by the evolving brain.

The capability for symbolic thinking and speech gave *Homo sapiens* the tools for inventing tales about mysterious and supernatural beings and sharing them orally within the tribe. The tales told of these gods became memes[72] that were shared with subsequent generations. The sharing of such tales by everyone in the tribe added cohesion to the group.

---

72. A term invented by Richard Dawkins that means: "Any unit of cultural information, such as a practice or idea, that is transmitted verbally or by repeated action from one mind to another in a comparable way to the transmission of genes." Wiktionary: The Free Dictionary.

Durkheim[73] has noted that common symbols, tales, and social laws served to unite a community and increase their survival odds.[74]

Gods Appear

Once upon a time – there were no gods. This was the case for our closest ancestor, the chimpanzee. Several tests of intelligence, comprehension, and cognition have been given to the chimpanzee, and although chimps have been found to have the cognitive abilities of a two-year-old child, there is no indication that they have the mental ability or language capability to think of or communicate the concept of a god. But man does have god concepts, so questions to answer are when did god appear over the seven-million-year period of evolution from chimp to man and why did our ancestors conceive of gods?

Our ancestors encountered things outside their experience, knowledge, or control (birth, death, lightning, seasons, floods, etc.), and when their language ability and cognitive capabilities allowed, they began to think about what was causing the many puzzling events around them. Since they had no base of knowledge from which to infer how these sorts of things came about, they thought of them as mysteries caused by something beyond their natural world—in effect, supernatural events. These early beliefs in the supernatural were applied to various imposing objects (sacred mountains, sacred volcanos, sacred rivers, etc.). This belief is known as animism[75]—the attribution of life to the inanimate—and it is considered the basis for the first religions, albeit primitive ones. Emile Durkheim defined religion as a unified system of beliefs and practices relative to sacred things.

As his cognitive capabilities expanded, *Homo sapiens* slowly replaced concepts of sacred objects with concepts of sacred supernatural humanlike forms as the cause of the mysteries, and the worshiping of humanlike gods became the norm. These appeared as polytheistic gods around 30,000 years ago in many guises (fertility gods, love goddesses, hunting gods, wine gods, rain gods, sun gods, etc.), each reflecting a mystery encountered in life. The polytheistic god concepts were also

---

73. Émile Durkheim, *The Elementary Forms of Religious Life,* Oxford University Press, 2001.
74. Ara Norenzayan and Azim F. Shariff, *The Origin and Evolution of Religious Prosociality,* Science, vol. 322, issue 5898, 2008.
75. Animism is the belief that all things are animated with spirits; it is the world's oldest form of religion.

considered anthropomorphic; that is, the gods had human attributes. For the first time, man was interacting with the concept of gods of humanlike form with supernatural powers.

About 20,000 years ago, the *Homo sapiens* brain evolved an executive planning capability. This was reflected in the transition from hunting and gathering to farming and in the domestication of animals and plants on a widespread basis. This capability was necessary with the emergence of larger settlements. In turn, as settlements grew, urbanization led to city-states that required governance with centralized leadership, authority, and governing laws. Leaders and kings increased their power by coupling their secular power with religious power that combined the laws of the group (state) and the group's religion (god).

With time, some gods gained favor with believers in relation to lesser gods and were elevated to be the most powerful of the polytheistic gods. This was the first step in the transition to monotheism, where one god was given the attributes of all gods. With one powerful god, monotheism aided in the solidification of the two major social concepts and their rules that had been made by man—states (groups) and gods (religion). Each had a set of rules and laws: secular humanistic societal laws emerged from observations of human activities and religious laws were developed by religious leaders to enhance the image of their gods and define the rules for worshipping them.

Laws adopted by emerging city-states were a mixture of secular social laws that had been conceived by almost every society and religious laws conceived for their specific god. The secular social laws—such as the Golden Rule ("do unto others as you would have them do unto you") and "thou shall not murder"—were common humanistic laws of morality conceived by many societies around the world. Religious laws were not part of the common morality, for they varied from religion to religion—and there were many—and were devoted specifically to a god. Secular social laws were the moral imperative of the group and reflected the common humanity of the society (how people lived with their neighbors), and religious laws reflected their wishes not satisfied by society (how to receive help in this life and beyond this life).

Urbanization led to the construction of holy places, such as churches, in which societies could gather and worship their spirits and gods. These places were important in enhancing socialization in a community. Significant structures for religious practices began appearing around 14,000 BCE. An example of an early polytheistic site is the Göbekli Tepe site in Turkey, which dates back to 10,000 BCE. And it was not alone, for other religious sites also appeared in the Middle East and Europe about that time. From then the more famous structures—the ziggurats in Mesopotamia, pyramids in Egypt, and stone circles in England (such as Stonehenge)—followed.

The city–state of Sumer rose in Mesopotamia about 6000 BCE, and it was in Sumer that the first literate civilization appeared, with recorded laws of the state and religious laws of their polytheistic gods. The Sumerian gods were held responsible for all matters pertaining to one's life. The first god about which there are records is Anu, the supreme god of water, knowledge, and creation. He was considered more powerful than all other gods[76] (the wind god, the sun god, etc.). The gods played central roles in the city–state and had political, judicial, and social administrative functions. Subsequent generations of believers upgraded or downgraded the importance of particular gods, redefined gods, and selected others. In addition to Sumer, there were other emerging city–states, and each had different gods and laws.

Sumerian religious narratives were sufficiently appealing to be worshipped by following generations; in essence, their godly narrative became robust enough to survive competition from other spirit concepts from year to year. Early Sumerian religious narratives became known from translations of the writings on clay cuneiform tablets discovered in Sumer and later in other city–states.

By around 2000 BCE, urbanization and proto-religions had also evolved in many other places, from the Indus Valley in India to the Nile in Egypt and rivers in Mesopotamia. In the Levant region of the Middle East, the Canaanites practiced polytheism that included the gods El and Baal. Yahweh was the national god of Israel and Judea.

As societies grew larger and rulers more powerful, there was a transition from polytheism to monotheism, a concept featuring a single all-powerful god. By this time, religions had evolved several major components: gods, narratives describing the gods, rules or laws (creeds) for worshipping their gods, a religious organization (priests) to manage the religion, and a meeting place for worshipping the god(s).

The concept of one central god, or monotheism, gave rulers a more powerful status and identity than they had had with the many less powerful gods of polytheism. This transition first happened in Egypt around 1330 BCE when King Akhenaton effected a shift in Egyptian polytheism by declaring Aten, the sun god, to be the main god to worship.

---

76. Robert Wright, *The Evolution of God*, Little, Brown and Company, 2009.

There were still other gods, but Aten was given status above them. The next generation reverted to polytheism, but later, polytheism declined, and monotheism advanced in Egypt.

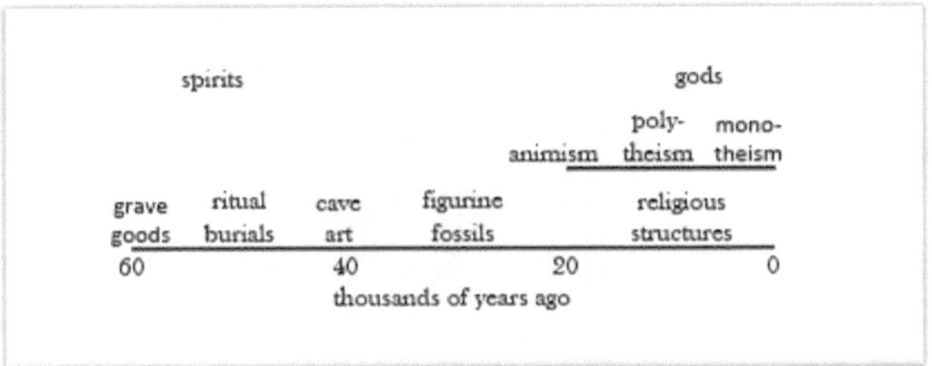

Figure 14 Evolution of Awareness of Spirits and Gods

The religious narratives and their codes of conduct and laws were transmitted orally until the invention of cuneiform writing on clay tablets. The cuneiform language for writing was created around 3200 BCE. Stories about kings, spirits, and gods were inscribed on clay tablets and widely distributed. An example is the *Epic of Gilgamesh*, written on clay tablets (Figure 15) about 2100 BCE to tell the story of a king who was elevated after death, because of his battles with ogres, to the status of a god. Epic poems incorporated the narratives of the gods and local myths.

Cuneiform writing became widespread and was used by kings to communicate the laws of the land. King Hammurabi developed his code of law while he was king. It was highly regarded and was used by others after his death in 1750 BCE. An example is a stone stele dating about 1600 BCE that lists the Code of Hammurabi, the laws of the land. The Code contains 282 laws and is an early example of a state's laws and religious laws. Also included in it is the claim that it was handed to King Hammurabi[77] by the chief god Shamash.

The culture in Babylonia at that time was reflected in this early law code. Justice was allocated to three classes of Babylonian society: property owners, freemen, and slaves. These laws served in the administration of justice and assisted commerce throughout the kingdom. Other kingdoms with other gods and laws have evolved, but the

---

77. See Appendix B for examples of Hammurabi's laws and their subsequent appearance in the Ten Commandments.

focus of these discussions will be on the evolution of the God of Judaism and, subsequently, the Christian God.

Figure 15 Epic of Gilgamesh—Tablet, 2100 BCE

Judaism's roots are as an evolving religion in the Middle East around 2000 BCE and it is considered one of the world's oldest religions. The name "Israel" first appears in nonbiblical sources in 1200 BCE on a stone honoring an Egyptian pharaoh. Judaism began as a tribal religion in the Levant region in the Middle East, and the god Yahweh was one of several "divine warriors" who had led a heavenly army against Israel's enemies. Other early Jewish gods included El (original god of Israel) and Baal (the god of storms). Jews transitioned from their earlier polytheistic gods to monotheism about 600 BCE upon returning from captivity in Babylonia when the royal court promoted Yahweh to be the only god of the kingdom of Israel and Judah and declared their god to be the creator of the universe and the one true God of all the world.

Monotheism has its advantages, but it created critical philosophical problems[78] for religions: it brought to light the question of how a good and all-knowing god can dispense both evil and good. In polytheism, good and evil were represented by separate gods who battled each other, but in monotheism, good and evil must come from the same source, the one god, presenting a problem for religions that remains unsettling today.

Judaism was the springboard from which Christianity evolved based on the teachings of Jesus, a Jewish rabbi and leader who died about 33

---

78. Lloyd Greering, *Reimagining God: The Faith Journey of a Modern Heretic*, Polebridge Press, 2014.

CE. Jews rejected Jesus as the embodiment of the coming of the Messiah, but the followers of Jesus elevated him to be the Son of God and a Messiah and founded a religion around him and his teachings. Jesus taught his message orally, often by using parables about life, but left no writings. Although his life was lived as a Jew, his life-example, and the messages he taught were collected by his disciples after his death and they serve as the bedrock of Christian theology.

Believers of the new Christian religion began documenting Jesus's life and teachings some fifty years after his death. As noted in the Bible, several versions of His messages are recorded by different people and at points differ. Over the next 400 years, the Christian Bible was assembled to describe their God and the life and morals of Jesus. The Bible incorporated much content from the Jewish Torah, including some of the earlier Jewish laws, such as the Ten Commandments, and supernatural concepts of God's creation of the heavens, Earth, and man (Adam and Eve).

Early writings that separated the Christian God from the Jewish God were first expressed in the writings of the apostles; an example is the writings of Paul around 50 CE, when he defined the Christian God and the divinity of Jesus. Thus, one can say that the Christian God appeared around 50 to 100 CE when first defined by Paul[79] and others. The description of the Christian God has been redefined[80] several times since then by councils of believers as the religion grew. With their God defined, other believers joined in writing and assembling the Christian Bible. The biblical scholar Richard Elliott Friedman[81] summarizes the assembly of the Bible:

> *The Bible is thus a synthesis of history and literature in harmony and sometimes in tension, but utterly inseparable.*

At the birth of Christianity during the first century of the Common Era, the ruling Roman Empire encompassed many religions, including Mithraism with its god Mithra. During the first two centuries, there was state-directed persecution of Christians in the Roman Empire while the pagans were left alone. However, Christianity survived those early days, and in the third century the Roman Emperor Constantine recognized

---

79. Paul, the Apostle, First Epistle to the Corinthians (54 CE).
80. For example, the Apostle's Creed of 390 CE further defines Christianity and God.
81. Richard Elliott Friedman, *Who Wrote the Bible?* New York: Simon & Schuster, 1987.

Christianity as a legal religion. By 380 CE he made it the state religion of the Roman Empire. He called the First Council of Nicaea to define Christianity, and the council produced the Nicene Creed that many Christians still use today. Constantine's support was a major boost for Christianity, for within twenty years pagan worship was prohibited in the Roman Empire and persecution of the pagans by Christians began throughout the Roman world. Christians followed the pattern of earlier religions: once in power, eliminate the rival religions, for yours is the only "true religion" and no competition is to be allowed.

There are many examples of pagan persecution by Christians. The death of Hypatia, a scientist and leading intellectual in Alexandria in 415 CE, is one example.

Although many people converted to Christianity, many cities were hesitant to give up polytheistic gods who were local favorites, so they called them saints and folded them into their Christian beliefs. Some were recognized by the church, while others remained folk saints to a region. An example was the city of Padua in Italy, which brought along their pagan god Anthony, made him a saint, and continued to address him in church to help find lost things. Other regions also held to their pagan favorite; one such was Francis of Assisi, who was elevated to sainthood by the church.

Jewish writings of the Old Testament and Christian writings of the New Testament were combined. The God concept, religious laws, and secular moral laws of the time were folded into the Christian narrative that has been remarkably important for millions of believers. As with other gods, the Christian God and narrative has brought feelings of comfort, support, salvation, and even awe to many believers. The narrative also contains cultural mores, symbols, practices, and laws that have contributed to our common humanity. David DeSteno[82] notes:

> *I have found that religious ideas about human behavior and how to influence it, though never worthy of blind embrace, are sometimes vindicated by scientific examination.*

The authority of Christian laws is based on the church's interpretations of its religious narrative, the Bible, a document written by believers. The growth of Christianity from the teachings of Jesus and subsequent

---

82. David DeSteno, *What Science Can Learn from Religion*, New York Times, February 1, 2019.

writings by apostles and others attests to the appeal of the Christian narrative. The Christian narrative also has a dark side; it incorporates a supernatural malevolent-entity concept (Satan). Such a concept had been discussed in earlier religions, such as Zoroastrianism. Biblical writers included such an entity in discussions of good and evil[83] to separate the good God from the evil spirit, Satan.

Hundreds of god narratives have been invented by man since *Homo sapiens* evolved the mental capacity to conceive god concepts. The Christian narrative, with its God, has become one of the leading religions in the world. However, there are many religions with many gods invented by man, and conflicts arise over which god concept is the "real" god. This is a difficult question for believers of these many religions to answer, for how does one select one god narrative over the others without proof of its authority and superiority when all gods emerged from ancient narratives written by different believers in the past? On the other hand, to believers in a specific god, the question is not important, for they have already entered into a belief agreement with their god. Regardless of what other religions may offer, one either believes or does not believe in one's god.

Philosophers have questioned the existence of gods in the past. For example, in 54 BCE, before Christianity entered the picture, Lucretius, a Roman philosopher,[84] took the measure of the early spirits and gods and noted:

> *There are no angels, demons, or ghosts. Immaterial spirits of any kind do not exist. The creatures with which the Greek and Roman imagination populated the world—Fates, harpies, daemons, genii, nymphs, satyrs, dryads, celestial messengers, and the spirits of the dead—are entirely unreal.*

Despite such views, over the years, the number of religious believers in gods has grown. Their beliefs in powerful supernatural entities have led to assigning the authorship of narratives to their gods. Since the narratives include morality, they fill a human need for structured support and guidance.

For some religious believers, personal revelations of their gods happen. They may be helpful to some, even awesome for the individual believer,

---

83. Elaine Pagels, *The Origin of Satan*, Random House, 1995.
84. Lucretius, *The Nature of Things*, trans. Martin Ferguson Smith, Simon & Brown, 2016.

but they are only exercises in creative internal dialogues in the brains of those who experience them, and they lack supporting proofs from external sources inasmuch as subjective experiences are unique to the one person.

However helpful they are to individuals, these beliefs are based on supernatural narratives and include actions that interact with and are in conflict with Nature's laws. Simply said, actions based on supernatural laws are not compatible with Nature's laws. Thus, when God's religious actions are placed over the laws of Nature or over the laws of the state, there are conflicts. Several examples of these conflicts are discussed next in the context of the Christian narrative.

VII. Conflicts

The history of conflicts between the Christian church's (God's) laws and commandments with Nature's (science's) laws and the state's (government's) laws is long and tangled. Before the scientific revolution and the Enlightenment in the West began around 1500 to 1600, the Christian church was the ruling theocratic power in most states and powerful enough to ignore conflicts with Nature's laws and punish scientists when they strayed from the dogma (lock them up or burn them at the stake). Although some clergy from time to time recognized the importance of natural science and made peace with it, the majority stuck to the older church dogma that conflicted with science and placed church dogma over new discoveries of science.

As discussed, early in the scientific age, the scientist Galileo was condemned by the church in Rome for his scientific discoveries that conflicted with the Christian dogma that Earth was the center of the universe. His defense—trying to be rational and separate secular science from the religious dogma demanded by the Vatican—failed. The church would not listen; why should it when it controlled both religion and the state?

This imposition of religious dogma on science by churches and the labeling of scientists as devils and workers of Satan have continued and have had varying degrees of success, depending on the social power of the church in society and in government at the time. In addition, governments have used religious laws to support their political power to reject scientific results for political gain.

Here are a few examples of conflicts of religion with science and the state over the last 2,000 years:

399 BCE. In Greece, the philosopher Socrates was killed by poison ordered by the Athenian assembly for corrupting minds and fostering nonbelief in the state's gods. In this case, the gods were pagan and were used by the state as a base for its exercise of power.

415 CE. In Egypt, Hypatia, a leading scientist, mathematician, and teacher in Alexandria, was murdered by a gang of Christian monks for being a pagan. Although a pagan, she was a leading teacher of students of all religions and had been honored by students of all religions. Her popularity made her an enemy of the church.

1536. In England, William Tyndale was executed, and his body was burned at the stake by the king of England for opposing the church's position on God and for writing an unapproved new translation of the Protestant Bible.

1600. In Italy, Giordano Bruno was burned at the stake by the Vatican for proposing scientific theories not approved by the Vatican.

1633. In Italy, Galileo Galilei was convicted of supporting heliocentrism, a scientific theory that the Vatican found to be "foolish and absurd" and heretical since it explicitly contradicted the Vatican's dogma of a geocentric universe. He was tried by the Vatican's Inquisition, found to be "vehemently suspect of heresy," and forced to recant. He spent the rest of his life under house arrest.

1834. In the United States, Abner Kneeland, a citizen of Massachusetts, was charged with blasphemy for saying he did not believe in (the Christian) God. He was convicted and served sixty days in jail. Blasphemy was a crime under the colonial charter of Massachusetts.

The publication of Darwin's theory on natural selection in 1859 ushered in a new wave of science's conflicts with Christians, for the evolution of man, as described by Darwin, does not need a cornerstone of Christian belief, the supernatural God, for creation of man. Some Christians reject the science and argue that there is an intelligent designer, their God, who created the universe, Earth, and man (Adam and Eve), all fully formed over a few days. Darwin's theory says that man has evolved over a much longer time from many ancestors. From scientific discoveries, we now know that biological life on Earth has evolved over a long period—3.7 billion years.

To scientists the rejection of Darwin's theory by fundamentalist Christians has been painful, as it is a direct rejection of facts that have come from many fossil discoveries linking man to our many ancestors along our evolutionary tree of life. From worms to *Homo sapiens*. Further, it is also a rejection of the molecular DNA evidence that has more recently been obtained from fossilized bones of our ancestors, such as those of chimpanzees, bonobos, Neanderthals, and Denisovans.

However, such conflicts are to be expected when a supernatural religious narrative[85] written some 2,000 years ago when man's

---

85. Judith Hayes, *In God We Trust: But Which One?* Freedom from Religion Foundation, 1996.

understanding of Nature was very primitive is used to replace proven scientific theories. Our understanding of Nature has expanded due to the many discoveries by scientists about Nature. The advance of scientific knowledge, when compared with the old, unchanging Christian narrative of creation by God, leads to conflicts.

Over time, Christians have faced many changes, not only in evolutionary science but also in the culture and social mores underpinning social governance. This widens the knowledge gap between the church dogmas not only with the scientific community, but also with society at large. Other explanations for the creation of the universe and man are found in religions, and all are based on supernatural creation narratives, each different from the others and each conflicting with Nature's scientific theories.

There is a role for religious narratives that give comfort to millions in time of need and advocate aid for the needy. When these stories are told in the context of a religious discipline, there are no conflicts for they remain in the supernatural world of belief. However, when religious beliefs and laws are imposed on others outside of the religious discipline, conflicts are inevitable.

Examples of three classes of conflicts are summarized: Nature and God, State and God, and Nature and State.

Nature and God

Nature's laws are often challenged by the religious. One conflict between religion and science observed by the author was particularly memorable. A student of mine invited me to his church (an evangelical mega-church) to hear his pastor talk about human creation. I accepted. On the prescribed Sunday, my student had arranged for a first-row seat in this large mega-church. After introductory music, the pastor came out and opened his sermon in a booming voice:

> "Was your daddy a gorilla? Those pointy-headed professors up on the hill[86] think that, so I will tell you the real story about God's creation."

I was surrounded by two thousand believers, so I decided it was prudent to not raise my hand and answer the question. The next day in class up on the hill, the pointy-headed professor did discuss Darwinian

---

86. He was pointing to the university where I was teaching.

evolution openly in a class with students holding different beliefs, including adherents to several Christian denominations, Jews, Muslims, and Atheists. The result of discussions by the class was an understanding of the role of supernatural stories: when confined to religious narratives being used internally, there is no conflict. But when supernatural stories are applied external to the religious environment in the world of science there are conflicts, for supernatural stories are not replacements for science theories. This open discussion in an academic setting is unfortunately not repeated in society at large very often.

The supernatural creation story of man with his soul, sins, and morality is central to many conflicts between religion and science. Examples of these conflicts and religious theories employed by Christians against science are outlined.

Creation

The Christian view of the creation of the first humans, Adam and Eve, is based on the biblical narrative in the book of Genesis. Two versions of the creation of the universe and man and woman are presented. Which one to select?

In the first chapter of Genesis is the following:

> *1:1 In the beginning God created the heavens and the Earth. [Man and woman are created a few days later.]*
> *1:23 And the evening and the morning were the fifth day.*
> *1:26 And God said, let us make man in our image, after our likeness.*
> *1:27 So God created man in his own image, in the image of God created he him; male and female created he them.*

In the second chapter of Genesis, God creates man and woman in another manner:

> *2:7 And the Lord God formed man of the dust of the ground and breathed into his nostrils the breath of life; and man became a living soul.*
> *2:8 And the Lord God planted a garden eastward in Eden; and there he put the man whom he had formed.*
> *22 And the rib, which the Lord God had taken from man, made he a woman, and brought her unto the man.*

As mentioned, Pope Pius XII once tried to argue that the big bang was proof of God making the universe but was talked out of it by Georges

Lemaitre, who was not only a Catholic priest but also a scientist. Since then, others have insisted on God having created the big bang, saying that only God could have designed such a fine-tuned event.

The timeline for the biblical narrative (Figure 16) of God's creations (the heavens, Earth, and Adam and Eve) has been estimated by Bishop John Ussher[87] to have begun in 4022 BCE. God is not mentioned before that date. The big bang is not mentioned, so for Christians their God created the universe, man, and everything else about 6,000 years ago.

The timeline for creation of the universe and man from the book of Genesis is based on words that were written when knowledge of the universe was limited by what could be observed with the unaided eye looking into the sky and people had no knowledge of biological evolution. Although the Christian supernatural creation narrative conflicts with Nature's dates, supernatural tales, such as Adam and Eve being the first humans and a talking serpent having influenced them, have been useful to the Christian narratives. As would be expected, there is a marked difference in the dates for creation between the biblical narrative and science. Knowing that the big bang occurred 13.72 billion years ago, Christians arguing for Adam and Eve 6,000 years ago raises a question: what was going on all those years in between?

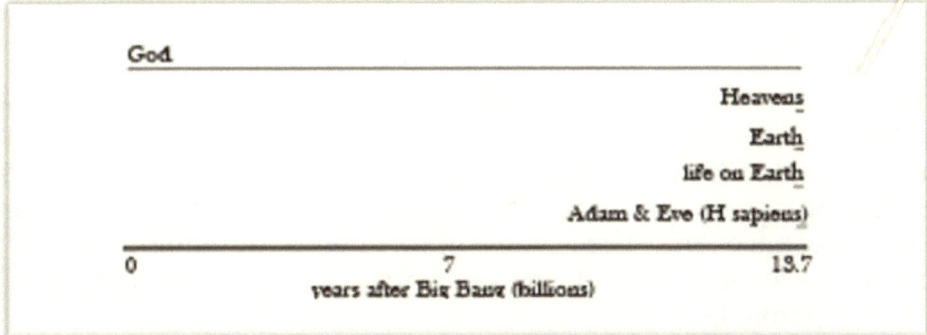

Figure 16 Creation of the Universe—God's Timeline

xyz By comparison, the timeline for Nature's creation of the universe is summarized in Figure 17. The big bang is the genesis of the universe at time zero (13.72 billion years ago). From that moment it took billions of years to spawn the expanding universe from which our solar system with its Earth were formed after 9.2 billion years. It took life on Earth 700 million years to evolve, and it took another 3.7 billion years of biological

---

87. Bishop James Ussher (about 1650) estimated the creation date of everything by God to be October 23, 4004 BCE.

evolution on Earth for the species *Homo sapiens* to acquire the mental awareness and cognitive capabilities to invent god concepts.

Had the Vatican accepted Galileo's plea 500 years ago—*the Bible tells us how to go to heaven, not how the heavens go*—Christians could use their supernatural tales in church services and scientists could have their theories of the universe in science activities, and many conflicts could be avoided. But that did not happen, and many Christians since Galileo's day

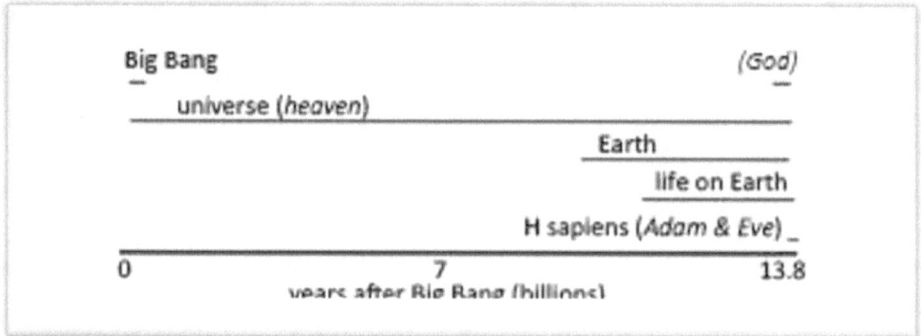

Figure 17 Creation of the Universe—Nature's Timeline

have continued to not separate the two, so the conflicts over creation continue.

Birth and Death

In the Christian narrative, God created not only Adam and Eve and a serpent but also every person after them. Science explains birth as a biological process stemming from the impregnation of a female egg by the sperm from a male. This is a process that took millions of years to evolve in our early ancestors. The science of birth is so overwhelming that most religious believers accept it but say, with no supporting data, that their God is responsible.[88]

This conflict on the individual scale causes big problems on a macro scale. The Christian narrative with God in control places no limits on the number of births. This is true in other religions also. It is a command from god that encourages more births. Authors of the religious narratives knew that more births made their tribe stronger and able to better survive in the struggles with other tribes with other gods.

This religious logic of birth has led to an unbridled increase in the population of the Earth to over 7 billion and with it has come the problems

---

88. Christine Tamir, Aidan Connaughton, and Ariana Salazar, *The Great God Divide*, Pew Research Center, 2020.

of hunger for many and a strain on the environment to feed them. Further, with the continued warming of the environment we are seeing a worsening of the problem. The increased world population is a major problem today and will be an even worse problem in the future with religions saying their god wants you to breed more.

Soul

The concept of a soul has varied in Judeo–Christian writings. The word "soul" in Hebrew is the same as the one for "breath," so the original meaning is "that which was the essence of a living body." Later, the Catholics gave a more physical definition:

> *The soul is a living being without a body, having reason and free will.*

These definitions of a soul are, to scientists, supernatural metaphors. Science removes the supernatural and assigns the concepts of a soul to natural functions of the brain. Francis Crick has argued that a person's unique identity, in effect his soul, is the subconscious output of the information stored and processed by his brain:

> *You, your joys and sorrows, your memories and ambitions, your sense of personal identity and free will, are in fact no more than the behavior of a vast assembly of nerve cells and their associated molecules. As Lewis Carroll's Alice might have phrased it: "You're nothing but a pack of neurons."*

The philosopher Paul Churchland summarily notes,[89]

> *[The brain] is the engine of reason and the seat of the soul.*

The old Hebrew meaning, "that which was the essence of a body," has been replaced with today's science meaning, "that which is the essence of a living brain."

Sin

The biblical Genesis story describes sin arising from the actions of Adam and Eve; led astray by a talking serpent in the Garden of Eden, they defied God's command and ate a forbidden fruit. With this act of disobedience to God, they committed a sin (the original sin) and were

---

89. Paul M. Churchland, *The Engine of Reason, the Seat of the Soul*, MIT Press, 1995.

banished from the garden into the world. The inclusion of a talking serpent and a man and woman who were created, whole, without evolution of the species (also true of the serpent) places this story in the supernatural realm. Sin, for Christians, is breaking God's commands described in the biblical narrative.

The scientific view of Darwinian evolution causes difficulty with this view, for it rejects the creation of fully formed humans and talking serpents within a few days after creation of the universe. If the story of Adam and Eve in the Garden of Eden defying God is rejected, then there can be no original sin, and without the original sin, the question of why God is needed to save man arises. Nevertheless, Christians continue to use the supernatural sin story to justify sins of all humans.

Man commits disobedience of laws, of course—not only biblical laws but also secular state laws, which are social laws written by man that are common to many societies. The Christian commandments, such as "thou shall not steal," are also secular social laws, which if broken by Christians are sinful, but which, if broken by non-Christians, are considered not sins but crimes.

Breaking a religious law is a sin only to believers. Breaking a state law is a crime to be judged and punished under state laws. Breaking a law of Nature is not a sin, but an act that results in something not working, for our universe operates under Nature's laws.

Morality

Charles Darwin in *The Descent of Man* argued that humans and nonhumans have much in common. Research on our ancestors, the primates, has shown that there is commonality in many areas. The growth of social interactions forms a beginning of our common morality, from primates to humans. Building on the biology of the evolving brain, Patricia Churchland[90] outlines the natural evolution of morals:

> *Moral values are rooted in a behavior common to all animals, the caring for offsprings. The evolved structure, processes and chemistry of the brain incline humans to strive not only for self-preservation, but for the well-being of allied selves, first offsprings, then mates, kin, and so on in wider insider "caring circles." Separation and exclusion cause pain, and the company of love-ones causes pleasure:*

---

90. Patricia S. Churchland, *Braintrust: What Neuroscience Tells Us about Morality*, Princeton University Press, 2011.

> *responding to feelings of social pain and pleasure brains adjust their circuitry to local customs. In this way care is apportioned, conscious molded and moral institutions instilled.*

Churchland also argues that biological evolution has produced social rules that constitute proto-morals. E. O. Wilson, an evolutionary biologist, argues that evolution has passed the making of morality through evolutionary biology:

> *Within biology itself, the key to the mystery is the force that lifted pre-human social behavior to the human level. The candidate in my judgement is the multi-level selection by which hereditary social behavior improves the competitive ability not only of individuals within groups, but among groups as well.*

As tribal size increased, shamans rose in power with promises of communicating with spirits that would protect believers from bad things happening. These human interactions with spirits and later gods helped group cohesiveness by providing guidance from common figures of power and authority (gods). As gods were conceived, religious leaders assumed the mantle of being earthly representatives of gods that gave them the religious authority to judge men on their adherence to religious rules and commandments that believers must follow to receive the gods' rewards. Since the religious narratives were invented by man, they reflected the social mores, culture, and accumulated moral rules of the time. Further, the fear of punishments and hopes for rewards dispensed by religious leaders gave them great power and authority to define morality within the tribe and later within the larger populations of states.

Many contemporary philosophers, such as Patricia Churchland and Scott Hestevold,[91] have argued that there can be morality without a supernatural god. The more religious philosophers say that morality is based on god's words. But the sequence of the evolution of human morality argues otherwise; *Homo sapiens'* morals in tribes existed before religious narratives were written. Later, men wrote religious narratives and attributed the morality in the narratives to gods they had conceived.

In the narratives some morals are in the form of religious commandments. In addition to religious laws, there are secular social

---

91. Louis W. Perry, *Jefferson's Scissors* (includes Scott Hestevold's Appendix, *God and Morality: Is There Any Relation between God and Morality?*"), 2009.

laws in the narratives that are based on observations of the conduct of individuals or groups that added value to the social health of the tribe. As noted, examples of secular laws are found in the Christian Ten Commandments, such as "thou shall not steal" and "thou shall not murder." Stealing and murder are destructive social activities within any group that, if allowed to go uncontested, will raise internal conflicts among members and diminish the group's survival probability. So, many societies as well as religions use these secular social rules to keep peace and order. Six of the Ten Commandments (see Appendix B) are secular social commandments[92] that are also found in many other religions and societies and in the sayings of philosophers, such as Buddha and Confucius. These socially derived laws that address the human social condition do not conflict with Nature's laws, for they are the results of observations by man. Religion-based laws, such as "honor thy God," are religion-specific based on religious narratives and are only for believers.

Flood

An example of conflicts about a physical event described in Christian narrative is the biblical worldwide flood. Such a flood was sent by God on the biblical date of 2348 BCE to cleanse the Earth (that is, kill all people and animals except Noah, his family, and a collection of animals) because of humanity's misdeeds. God declared that humanity was to be saved from the flood and start over with a few animals and a family (Noah's) by floating in a preconstructed ark. Biblical scholars point out that there were two biblical writers providing different background stories for the flood, two different sets of instructions about the animals, two different time frames for the flood, and two different stories about Noah and the animals leaving the ark. Since God's narrative does not pick one, you get to pick one.

Science has not been able to find any evidence of such a worldwide flood. A flood of the Earth in forty days would require a rainfall of about 250 inches per hour, all day and all night for those forty days, in every place on Earth. There is no weather source known on Earth that could supply such a rainfall, but supernatural stories do not need to obey Nature's laws.

---

92. Honor one's parents and refrain from murder, adultery, stealing, and giving false testimony.

Intelligent Design

Some Christians attempt to replace Darwin's theory with various religious pseudo-theories. One such theory is the intelligent design theory. The evolution of humans described in Chapter V is rejected by many Christians with the argument that the human body is too complicated to have evolved by Nature as described by Darwin's theory of natural selection. Only a supernatural intelligent designer (in effect, God) could have designed humans, they say. Two examples of biological complexity cited by intelligent design theory advocates as proofs include the human eye and the blood-clotting system.

To accept the intelligent design pseudo-theory for the creation of the biblical Adam and Eve requires the rejection of discoveries of billions of years of natural evolution. One example is the thousands of fossils of animals that have been found. DNA tests and analyses that have been done across fossils of many species have only further supported Darwinian evolution.

As for human design being too complex, an example of the science of the evolution of blood clotting in vertebrates has been described in detail by Russell Doolittle,[93] an evolutionary biologist, and others. It describes the step-by-step evolution of blood clotting over 400 million years across many species, from early vertebrate fish to humans. The details of the many evolutionary changes by species over this time are complicated but explainable in terms of the biology we know, without reference to a god.

As discussed, *Homo sapiens* interbred with *Homo neanderthalensis* about 50,000 years ago and probably at other times. Although the Neanderthal species became extinct about 30,000 years ago, its genes continued to survive in certain *Homo sapiens* populations with DNA levels of Neanderthal genes up to 4 percent. One would be hard-pressed to find an argument to support an intelligent designer (God) inserting pre-human (Neanderthal) genes, which Christians say never existed, into Adam and Eve. The intelligent design theory is an unsupported supernatural concept that is rejected by the scientific community.

Status

Christians of all types and theists alike can respect and endorse the social messages by Christians about man's common humanity, such as those given in the parables and life examples of Jesus. Atheists argue that

---

93. Russell F. Doolittle, *Evolution of Vertebrate Blood Clotting,* University Science Books, 2013.

there are no supernatural gods but agree that there can be a spirit of common humanity in societies that can be embraced by all.

Thomas Jefferson recognized this when he edited the Bible's New Testament and from it assembled a little book now known as the *Jefferson Bible*[94] that he constructed by removing passages that invoked the supernatural, such as the supernatural birth and death of Jesus, but leaving the life of Jesus, his parables, and other humanistic parts of Jesus's message. Jefferson attempted to illustrate the value of both religion's passion and common humanity described by Jesus and science's reasoning for common humanity. Religious believers are free to believe in supernatural gods, but it is not acceptable for the religious to deny the understanding of scientific discoveries about Nature.

However, many believers have difficulty holding a rational view of Nature in light of the passion they have for their religion. The philosopher Jonathan Haidt[95] noted:

Reason is the servant of passion.

Where there could be a non-confrontational relationship of nature, state, and god, religious passion is often a barrier. The rationalist view of science and the Constitution is being severely tested in our country today. One example is the anti-abortion position that denies the science of conception and birth. Another example is the rejection of the science of environmental climate change by deniers of worldwide weather changes from CO_2 accumulation in the atmosphere caused by man's activities, such as burning coal.

State and God

Early tribal leadership was a combination of shamans and warriors: shamans addressed people's concerns about the mysterious, offering supernatural answers by evoking local spirits, and warriors focused on defending the tribe and how they could enhance their ruling power. As the size of tribes grew, so did the need for communication with tribal members. Out of this need, formal structures and recorded laws came from the secular rulers and religious leaders. Although these leaders had a long, interwoven, contentious journey together, the struggle continued

---

94. Thomas Jefferson, *The Life and Morals of Jesus of Nazareth, Extracted Textually from the Gospels in Greek, Latin, French, and English,* Forgotten Books, 2017.
95. Jonathan Haidt, *Righteous Mind: Why Good People Are Divided by Politics and Religion,* Pantheon Books, 2012.

over who had the power and who was in charge. Many times, kings and religious leaders joined to exercise dual authority and power in theocracies.

After a thousand years of God-given theocratic rule in most countries of the Western world, it was in England that the first steps were taken to limit the power of kings, nobles, and dukes, which, they claimed, came from God. A rebellion by English nobles in 1215 forced the king to sign the Magna Carta, a document that limited some of the king's power over nobles, though not over all peoples. Even so, it has been described as the first of a series of instruments[96] recognized as special to the cause of advancing the rights of individuals in states.

Laws are social contracts, and some are the outcome of many steps in the struggle by citizens to acquire some rights in deciding under what laws they will be governed. The rise of the parliamentary form of government in England, pushed by efforts to expand the power of the people, continued the transition of power from the kings and clergy to the people. But change in state laws affecting the people was difficult to achieve, for it had to overcome centuries of the almost unlimited power in theocracies exercised by kings and clergy. It took many years for the people to have a powerful voice in their government. The philosopher Patricia Churchland[97] noted:

> *The social morality of the people is the basis for civil laws and critical for governance not as a divine business or a magical business, but as an essentially practical business. Making good laws and building good institutions, should be thought of as cooperative tasks requiring intelligence and understanding and a grasp of the reverent facts.*

Secular Constitution

Five hundred and fifty years after the Magna Carta (1789) our founding fathers wrote and signed the Constitution. It gave rights to all citizens—the powerful and the disempowered of all religions. The new nation had no entrenched history of monarchies or theocracies to overcome, only an occupying power, England, to cast off. Almost every citizen was an

---

96. The Habeas Corpus Act, the Petition of Right, the Bill of Rights, and the Act of Settlement were all major contributions to English common law, which Americans inherited.
97. Patricia S. Churchland, *Touching a Nerve: The Self as Brain*, W. W. Norton & Company, Inc., 2013.

immigrant or the son or daughter of an immigrant and most but not all were from different Protestant Christian denominations. The Constitution did not address the issue of slavery for blacks. It left the power to regulate slavery, as well as most powers, to the individual states.

This new democratic government was secular yet encompassed all religions. The founding fathers believed that the people (excluding slaves) should be the governing authority of the state and that religion should be separated from the government. The secular Constitution formed the basis of a government of the people, by the people, and for the people without the interference of religions.

In a democracy, the morals and values of the citizens of many religions are incorporated into state laws and then enacted by a majority. Out of this our common humanity is expressed in laws that are secular, for they serve all citizens from many different religions. Our country with its separation of church and state has been a haven for many who were religiously persecuted for heretical religious beliefs in other countries.

Thomas Jefferson gave us an insight into the separation of Nature's laws from biblical religious laws in government by arguing,

> *Fix Reason firmly in her seat, and call to her tribunal every fact, every opinion. Question with boldness even the existence of God; because if there be one, He must approve the homage of Reason rather than that of blindfolded Fear. Do not be frightened from this inquiry by any fear of its consequences. If it ends in a belief that there is no God, you will find incitements to virtue in the comfort and pleasantness you feel in its exercise and in the love of others which it will procure for you.*

Thomas Jefferson and James Madison argued for the separation of religion, first in their state of Virginia, and then with help made it a foundational concept of the new American democracy. They first toiled to make their state government in Virginia free of religious involvement and then proceeded to address the separation of church and state at the national level. The prevailing civil philosophy of the delegates writing the Constitution was a political stance premised on an understanding:

> *Chiefly the Protestant notion of religion was as a private assent to a set of propositional beliefs.*

At one time during the deliberations at the Constitutional Convention, a preamble describing the new country as essentially a Christian nation was discussed. It was placed before the delegates for consideration but failed to receive adequate support and was dropped. A secular preamble was subsequently written and approved and is the one we have today; the Constitution contains no reference to God[98]. Other Christian references were also rejected, and the concept of the separation of the church from the state was agreed on and codified in the First Amendment to the Constitution:

> *Congress shall make no law respecting an establishment of religion or prohibiting the free exercise thereof.*

Our founding fathers had a range of beliefs. At one end of the spectrum were the deists, who were not strong church supporters. But others, on the other end, were people like Joseph Priestley, founder of the first Unitarian Church in America. Priestly commented on the beliefs of one of the key founding fathers, Benjamin Franklin, a deist:

> *It is much to be lamented that a man of Dr. Franklin's general good character and great influence, should have been an unbeliever in Christianity, and also have done so much as he did to make other unbelievers.*

At that time, many Christian leaders were concerned with the exclusion of their God from the new government, and some Christians attacked the founding fathers for their lack of beliefs and for their views on the separation of God from the new government. A history textbook at the time from a Christian university labeled Thomas Jefferson a liar and an antichrist for his deistic beliefs:

> *American believers must be aware of his [Jefferson's] views of Christ as a good teacher, but not as the incarnate son of God. As the Apostle John said, Who is a liar but he that denieth that Jesus is the Christ? He is antichrist, that denieth the Father and the Son.*

The powerful Timothy Dwight, president of Yale and a leading Christian, attacked James Madison, who was later elected president, for his support of the Constitution that does not mention the Christian God:

> *The nation has offended Providence. We formed our Constitution without any acknowledgements of God,*

---

98. Issac Kramnick and R. Laurence Moore, *The Godless Constitution, The Case Against Religious Correctness*, Norton.

> *without any recognition of His mercies to us, as a people, of His government, or even His existence. The [Constitutional] Convention, by which it was formed, never even asked once, His direction, or His blessings, upon their labors. Thus, we commenced our national existence under the present system, without God.*

Other Christians predicted the failure of any state that did not include God in their governing message. Religious organizations had rarely acknowledged (who knows of any?) that a secular democracy would produce a government with more freedoms for its religious citizens than a Christian theocracy with one religion above all others would. And indeed, the Vatican was initially not supportive of the American secular democracy for its democratic precept of the separation of church and state and the forbidding of any religious test for public office. Even as late as 1864, Pope Pius IX declared in his *Syllabus of Errors* that

> the Church ought [not] to be separated from the State, and the State from the Church.

Theocratic governance has been a long tale of the repression of religious rights of other religious organizations. Theocratic culture and God's laws (not contestable in the eyes of the church) have greatly limited the transparency of government (no one can question God's representatives about crimes), and people have suffered. As in many organizations, including theocratic ones, there was no voice of the people, no visibility by the people—and no oversight power by the people leads to corrupt leadership. The worldwide sexual abuse of children by numbers of Catholic clergy is an example of what occurs when there are no checks by the secular state to enforce the laws.

The Constitution is a secular document designed to guarantee personal freedoms, including the freedom of religion for all citizens. It is a living document that has been changed from time to time with twenty-seven amendments, including changes in the definition of a citizen.

When our Constitution was written, slavery was acceptable and recognized as an engine of commerce in many states within the country. Its acceptance was challenged by a group of southern states, and a civil war was fought with those wanting to continue slavery. In 1864, during the US Civil War, heightened awareness of the evilness of slavery gave President Lincoln the opportunity to press for the Union to amend the Constitution (Amendment XIII) and abolish slavery. Although legally abolished, the culture of slavery in the form of holding peoples of color as inferior continued. Over the years many protests of actions by the government and police stem from an awareness and even acceptance of the old culture of slavery.

The culture of slavery has darkened the moral values of many Western countries and religions. At the founding of the Christian religion the Bible states that Hagar, the slave girl of Abraham, was beaten and cast out by Abraham's wife Sarah. There was slavery then and it has continued in the Western world, where slave trading was conducted by many countries with slaves taken from many countries. Muslims, Jews,[99] and Christians were slaves at one time and at other times slave traders. The Torah and the Talmud contain various rules about how to treat slaves. The Christian Bible still reads that slavery is acceptable to God.

Christian justifications of slave trading were made formal by the Catholic church. An example of justification of the enslavement of Muslims is given in the Vatican's canon law, the *Decretum Gratiani,* in the fourteenth century. After this time, the major slave trading was from Africa by English and French traders to the Americas. Hopefully, the protests by many young people in our country in 2020 will result in a change of the cultural perception of peoples of color and remove the vestiges of otherness left from slave trading.

The Constitution has also harbored vestiges of biblical morals that were contemporary in society, such as the denial of full human rights for women and homosexuals. Changes in the Constitution have been made to reflect our evolving culture through amendments granting voting rights for citizens—for people of any race and color with Amendment XV (1870) and for all people of any sex with Amendment XIX (1920). Other changes, such as rights for homosexual citizens, are still in flux. The Constitution reflects the slowly changing culture and morals of our society.

In general, our secular constitutional government has worked well, if slowly, and public education by the government has largely kept religion separate from schools while providing freedom for all religions. However, religious intrusion into the government waxes and wanes with variations in the elective strength of Christians in government. When Christians are in a majority, there is constant pressure to insert the Christian God and His laws into the government and at times these inroads have removed some of the wall separating religion from the government. The change of

---

99. Eli Faber, *Jews, Slaves, and the Slave Trade*, NYU Press, 1998.

the motto of the country from *E pluribus unum* to *In God We Trust* in 1956 is an example.

Christians pushing for a "Christian nation" do not seem to go away, for fundamentalist Christians still argue that the country would be more moral if it was a "Christian nation." This is a problem of maintaining a secular government in which no one religion is supported and all are treated equally. As many have said:

> *The price of freedom is eternal vigilance.*[100]

History gives us a little irony in relation to this argument. The Confederate States of America was a country formed from states that split from our country over the issue of slavery. Its constitution, although modeled on ours, was different, for it embraced slavery and appealed for guidance from the Christian God by

> *invoking the favor and guidance of Almighty God.*

History has shown that their "Christian nation" constitution gave them no help, for the outcome of the Civil War was that the secular country (the USA), without the guidance of any Almighty God, won the war. Such is the benefit of being a "Christian nation."

Landmark Court Cases

The theory that God's laws are always superior to secular laws is still put forward by some Christians, even in secular governments, when fundamentalists attempt to inject their religious views into the secular business of the state. Christians have long attempted to impose their Christian morality on citizens: for example, their views on abortion (the right of women to care for their bodies) and marriage (the right of men and women to choose their partners in marriage).

Two landmark lawsuits, one about abortion (*Roe v. Wade*, 1972) and one about marriage (*Obergefell v. Hodges*, 2014), made their way to the Supreme Court. In both cases, the Supreme Court ruled in favor of people deciding their actions rather than following religious dogma.

Roe v. Wade[101] – The Supreme Court in 1973 ruled, 7–2, that a right to privacy under the due process clause of the Fourteenth Amendment extended to a woman's decision to have an abortion, but that this right

---

100. A simplified variant of a line from a 1790 speech by Irish orator John Philpot Curran.
101. *Roe v. Wade*, 410 U.S. 113, 1973.

must be balanced against the state's interests in regulating abortions (protecting women's health and protecting the potentiality of human life).

Obergefell v. Hodges[102] – The Supreme Court in 2015 ruled, 5–4, that the fundamental right to marry was guaranteed to same-sex couples by both the due process clause and the equal protection clause of the Fourteenth Amendment to the United States Constitution. The Court required all states to issue marriage licenses to all couples and to recognize same-sex marriages validly performed in other jurisdictions.

Some countries with long histories of theocratic governance have continuing problems removing the church's authority from state governance. Ireland is an example of a state striving to establish the authority of its state over the unwholesome reach of the church into the civil protection of children. The public unfolding of the tragic Irish clergy pedophile cases has highlighted the reluctance of the Catholic church to give up on its long-held theocratic position of elevating canon law above civil law.

Over many centuries, the Catholic Church was an essential part of an Irish theocracy. Ireland was the most religious of all European countries—effectively, a Catholic theocracy. Sarah Lyall[45] summed up the situation:

> *This is still a country where abortion is against the law, where divorce became legal only in 1995, where the church runs more than 90 percent of the primary schools and where 87 percent of the population identifies itself as Catholic.*

The need for the government to be independent of religion finally became a glaring issue in Ireland after the church gave only lip service to protecting the civil rights of children. A *New York Times* report on the statements by Irish Prime Minister Kenny tells the story:

> *After 17 years of revolting revelations the latest report on the Cloyne diocese in County Cork exposed an attempt by the Holy See to frustrate an inquiry in a sovereign, democratic republic as little as three years ago, not three decades ago.*
>
> *The report, he said, excavates the dysfunction, disconnection, elitism, the narcissism that dominates the culture of the Vatican to this day. The rape and torture of*

---

102. *Obergefell v. Hodges*, 576 U.S. 644 (2015).

children were downplayed or managed to uphold, instead, the primacy of the institution, its power, standing and reputation.[103]

For any leader in the many countries scarred by pedophile priests to have pulled back the curtain to expose the profane amid the sacred would have been remarkable, but for the devoutly Catholic prime minister of a nation whose constitution once enshrined the special position of the church to do so was breathtaking.

Kenny's reported statement is a powerful one supporting the argument that theocracies and governments that mix religion and state do not provide good government for all of the people, and it emphasizes the need for the separation of church and state. The checks and balances provided by an independent government and science are essential. America included separation of church and state in its Constitution in 1778; Ireland is now beginning to follow the democratic example of separating the church from civil powers, but the extensive powers exercised by the past Catholic theocracy die slowly.

Nature and State

There have been cases where state governments have dictated to science and ignored Nature's laws. A notable instance of this occurred in the Soviet era in Russia during the 1930s, when the Stalinist government elevated Trofim Lysenko, who embraced a false (in the eyes of the scientific community) theory of genetics, to be the national scientist. A big problem soon became apparent when his theory was implemented and failed to increase farm production. This was during a critical time in Russian agriculture and people were dying for lack of food. Farming was government controlled. Lysenko's rejection of Mendelian genetics in favor of his pseudoscientific theory of genetics resulted in poor crop yields, which led to the deaths of millions. The authority of the state being used to reject the scientific community's views of Nature had tragic consequences.

In our own country today, the government has supported policies counter to Nature's laws, claiming that global warming is not caused by man's actions. In doing so, the government has rejected scientific theories and research data on global warming that show that it is caused by an

---

103. *Cloyne Report on the Diocese of Cloyne*, An Irish government document, Judge Yvonne Murphy, 2011.

excess of human-produced CO_2 produced mainly through the burning of fossil fuels. The resulting greenhouse effect in the atmosphere causes an increase in air and ocean temperatures, melting of ice around the globe, and dangerous weather conditions. Again, the authority of the state being used to reject the scientific community's views, which are based on the authority of Nature, has not worked out well.

Although we should honor Nature's laws, scientists do not always get it right. A classic example occurred in 1949 when Portuguese neurologist António Egas Moniz received the Nobel Prize for Physiology or Medicine for his development of the prefrontal lobotomy. He promoted the procedure by declaring it successful based on just ten days of postoperative data. Largely because of the publicity surrounding the award, other physicians recommended the procedure, in disregard of modern medical ethics. Favorable results were initially reported by reputable publications like the *New York Times*. It was estimated that around forty thousand lobotomies were performed in the United States before they were suspended after protests from the scientific community and the medical establishment that the procedure was not beneficial.

A textbook example—pun intended—of state power being used by a religion to reject Nature's theories was the case of Christian fundamentalists projecting their political power over the state to reject Darwin's theory in public schools. The state of Alabama blatantly acquiesced to a Christian majority of voters by passing a law requiring insertion in all high school biology textbooks of a paragraph stating that Darwin's theory of evolution is "only a theory." This rejection of science theories opened the door for biology teachers in the state to teach the Christian supernatural theory of the creation of Adam and Eve. The state had placed the authority of God's laws over those of Nature.

This view has been expressed in many books[104] by Christians who continued to argue that there is no conflict between their biblical text and Nature's laws if they are interpreted properly.[105] However, the scientific world sees Darwin's theory as the underlying explanation for the evolution of all biological life; dismissing its importance misinforms students about Nature, for Nature does not employ supernatural gods or

---

104. Two examples are *The Language of God: A Scientist Presents Evidence for Belief* by Francis Collins and *Coming to Peace with Science* by Darrel Falk.
105. See *Thank Evolution for God, The Role of Nature and God in Evolution*, Xlibris, Louis Perry, 2012.

do supernatural acts. Public education is a major civil commitment in our country, and doing it incorrectly is a disservice to all citizens.

Religious Pressure

Early in our history, some regions of our country acted as if it were a "Christian nation." Abner Kneeland, Massachusetts pioneer, freethinker, and Universalist minister, got in trouble in the 1830s for publishing an article[106] in a newspaper, the *Investigator*, in which he wrote,

> *Universalists believe in a god which I do not; but believe that their god, with all his moral attributes, (aside from Nature itself,) is nothing more than a chimera of their own imagination.*

Kneeland was charged with blasphemy in 1834 for saying this, endured three trials, was convicted on one count of blasphemy, and served sixty days in jail. The prosecuting attorney for the commonwealth of Massachusetts told the jury that if Kneeland was not punished,

> *[marriages will be] dissolved, prostitution made easy and safe, moral and religious restraints removed, property invaded, the foundations of society broken up, and property made common.*

Kneeland's appeal to the state supreme court concluded with a split verdict of guilty in 1838. It was the last case of blasphemy brought to a court of law in this country. As is to be expected in a democracy, the boundaries of the law have been and will be continually tested; this has certainly been the case with respect to the interface between religion and the state.

Many Christians assume that God has authority over science and that they should insert religious supernatural theories into public school science education (teach creationism[107] or intelligent design) and into government research (limit stem cell research). This is to be expected in the general US population, of which about a third (33 percent) believes in Darwin's theory of evolution, while 67 percent continue to believe in supernatural biblical involvement. Among evangelicals, 96 percent

---

106. *Abner Kneeland*, Freedom from Religion Foundation, 2011.
107. Creationism is the biblical version of God's creation of the heavens and man.

believe in supernatural involvement and 4 percent believe in Darwinism.[108]

The United States has one of the highest rates of Christian religiosity in the industrialized world, and this continues to be a barrier to the independence of science. The scientific community believes differently, with 80 percent of its members saying they do not believe in a god, but scientists are a small percentage of the population.

In many ways, Christians have resisted the democratic principle of the separation of religion from the government. For example, they have organized school-led prayer in classrooms and, in some places around the country, erected monuments of the Ten Commandments and Christian crosses on public land. Most of these acts have resulted in lawsuits to enforce the separation of church and state. One example, close to the author's home in California, is a Christian cross erected on public land atop a hill called Mount Soledad. Lawsuits and legal wrangling have continued for twenty-three years. The case was appealed to the Supreme Court, which declined to reconsider a lower court ruling against retention of the cross on public land. The separation of church and state was upheld, but Christians found another way to retain the cross on the hill: they bought the land from the government, making it private property, so the cross remains.

Another myth of the "Christian nation" is that our laws are based on God's commandments. This myth has been supported by many believers who obviously fail to notice that a number of biblical commandments are not used in our government. Two examples from Exodus 34:

> *No. 3 - Dedicate the first offspring from every womb to God.*
> *No. 10 - You shall not boil a young goat in its mother's milk.*

These do not seem like a sound basis for democratic laws in a diverse society today.

Our Constitution has been the ever-changing product of a long struggle by citizens who came from many countries to throw off the yoke of religious dogma. Although the country did incorporate some biblical law (for example, the inferiority of women) in the writing of the Constitution, much of that has been changed or is in the process of being changed. Episcopal Bishop Spong has observed some of the consequences of biblical law in a democracy:

---

108. Pew Research Center.

> *If we literalize the Scriptures, as Christians have tended to do and which fundamentalists do without apology or hesitancy, we also literalize the prejudices of that era, which were against democracy, against people of color, against women and against homosexual persons.*

One recent conflict between a religious authority and medicine in our secular country touched all three: religion, science, and democracy. In a Catholic-administered hospital was a pregnant patient with an emergency illness that threatened her life. The hospital's medical team assigned to save the woman's life said they would have to abort the fetus. The hospital's ethics council (which included a Catholic sister) met quickly and approved the procedure, for it gave doctors the best chance to save the woman's life, and an abortion was performed.

The local Catholic bishop heard the story and excommunicated the Catholic sister on the medical ethics team for violating the Vatican's ban on abortion. The hospital refused to fire the sister or the doctor who performed the operation, saying they acted ethically and humanely. Not getting his way about the firing of the sister or the doctor, the bishop then excommunicated the whole hospital, putting hundreds of patients, many of whom were Catholics, at risk. The hospital ignored the bishop and continued to provide medical services to the community, but it no longer had Catholic services within the hospital. In the end, the authority of the secular government's medical ethics guidelines was followed, and the authority of the church was rejected.

This is a clear case of religion overstepping its authority and trying to make medical decisions (what procedure to perform and who should get what treatment) that science should make. Sadly, this case undermines the major humanitarian services Catholics provide to thousands throughout the country. In some states, Catholic facilities have the highest percentage of the hospital beds in the state.

Having morals and laws given by "one's own special God" presents a problem. Patricia Churchland[109] noted that this problem has existed since Socrates' time:

> *For one thing, it [religion] makes a virtue of intolerance—those who disagree regarding a matter of morals must be dead wrong.*

---

109. Patricia Churchland, *Brain-Wise: Studies in Neurophilosophy*, Massachusetts Institute of Technology, 2002.

Moving beyond the intolerance of theocracies has taken a lot of time and a lot of philosophizing. Aristotle is but one philosopher among many, from Marcus Aurelius to David Hume, who have declared that divine guidance is not required for morality in government. Theocracies have never been a friend of freedom of secular laws, for they insist on the support of religious dogma and the canon law of one religion, excluding all others.

Theocracies come to power by the force of religious organizations in societies and practice intolerance by projecting the fear of God's vengeance on those opposing them. Christian theocracies grew and expanded in many Western countries and were the norm in the West for over a thousand years. But all large organizations in positions of power, including churches and governments, are fertile fields for corruption, and in churches from divine power came divine corruption.

The Catholic church has been a classic example of a large organization in power for centuries, and after many years of governing throughout Europe, Catholic theocracies descended into widespread corruption. An early form of corruption was the selling of indulgences (money paid to the church to get souls out of purgatory); that is, one could pay money to buy freedom in the afterlife from God's punishment of one's sins. The Catholic Church used this practice widely in the Middle Ages to extract money from various local states and their people. As the jingle of the time noted,

> *Coins in the coffers ring*
> *The soul from purgatory springs*

Indulgences were essentially a religious tax by the Vatican on believers, which they used to build cathedrals, the most lavish being St. Peter's in Rome.

Opposition to indulgences eventually became widespread, starting with the protests by Jan Hus, a Bohemian priest, in 1415. About a century later, in 1517, a protest by Martin Luther—a German priest and professor of theology—became a central church issue. He had been to Rome and had seen where the money from peasants in his part of Germany was being used, and he wanted to put an end to it. Protests demanding changes in the church went unheeded. These and other demands for change led many Catholics to split from the Vatican, causing a division that culminated in the Protestant Reformation. The sheer extent and force of the Protestant Reformation from the 1500s onward caused the Catholic church to react defensively by expanding its police force (the Inquisition) to handle internal threats to its Christian dogma and external threats from Protestant governments in many parts of Europe.

A response of change known as the Counter-Reformation to the Inquisition grew in power as the number of believers who were labeled heretics grew. At the local level, the mythos of Jesus continued as priests serviced the poor and did humanitarian work, but at higher levels of the church's organization, there was a secretive and kingly management structure increasingly separating the church bureaucracy from the people. The amount of effort required to defend the church against opposition had been highlighted by the Reformation, and internal corruption by the bureaucracy diluted the church's humanitarian efforts. To counter these problems, the bureaucracy continued to establish new church dogmas and enforce the old ones. In effect, Christianity had become a big old business plagued with all the systemic human problems of big organizations rife with corruption that we painfully see around us today.

During its history, there has been and continues to be a human struggle within the Christian church bureaucracy between those doing good work for the poor and those grabbing power for themselves. The church became more conservative and expended much of its energy upholding church dogma and resisting change. Holding on to its power became the dominant mission of church leaders. Those carrying out Jesus's humanitarian message had their adherents, but they were shortchanged by the upper reaches of the power structure in the church, who mainly directed their efforts to retaining the church's power base. But all human organizations, however powerful, are changed over time by emerging internal and external forces.

In the case of Christianity, change was driven by the inability of the church organization to effectively manage itself to serve its basic goals. The bureaucracy, which said its actions were by the authority of God, did not welcome feedback from the people they were serving. How can you contest the Word of God, you lowly peasants? With uncontestable power and wealth for over a thousand years, the church bureaucracy became corrupt and disconnected from its believers. Reform from inside proved ineffective, leaving many dissonant groups with only one choice— breaking away from the Vatican and forming their own churches.

The church's bureaucracy, consumed with internal problems, was also presented with external problems. It was only to be expected that with the steady increase of discoveries in natural science, secular philosophy, and governance, Christian dogma would come under attack by modernity.

With the printing press making information increasingly available after 1450 to scientists, philosophers, and the rising informed merchant class across Europe, pressure for change accelerated. Work by scientists and philosophers provided the information with which citizens started exposing, peeling back, and removing some of the many layers of theocratic dogma that had accumulated. With Christian theocracies losing much of their civil power after the Reformation, state laws installed by the people become increasingly secular. Finally, in democracies, the authority for determining bad (evil) persons in the context of society was removed from religious (God's) laws and given to secular laws enacted by the citizens. "Evil" acts, blasphemy, and other conduct against the Word of God were replaced by specific secular criminal acts.

A major hurdle for the Catholic church's modernization is its acceptance of the secular laws of host nations and recognition that the church's morals and laws do not extend beyond the church community. The argument that the church can police itself has proved to be false. There have been many past examples, but recently the exposure of worldwide pedophile crimes within the Catholic clergy has sadly and dramatically proved that the church is not capable of providing for the most vulnerable of its flock, children, and affording them basic human rights.

VIII. Summary

This book outlines what has been learned from studies of Nature's trajectory since its beginning with the big bang. What we would like to know is captured in Piet Hein's famous poem[110]:

I'D LIKE

I'd like to know
What this whole show
is all about
before it's out.

We have learned a little about Nature's laws that gave us the big bang and the expanding universe, the first self-replicating molecule (man's universal common ancestor), and later *Homo sapiens*. These have evolved under the laws of Nature, which allowed for the probabilities, however small, that those things could happen. The creation of the universe with the big bang may have been a very small-probability event, but it did happen. Some 9.2 billion years later, the evolution of biological life from the formation of amino acids and their arrangement into a self-replicating DNA molecule—also a small-probability event—did happen. From this beginning of life, vertebrates and *Homo sapiens* evolved; although that was also a very small-probability event, it also happened. What other small-probability events are in store for humankind?

The natural history of the evolution of man is special, as noted by Stephen Hawking:

> We are just an advanced breed of monkeys on a minor planet of a very average star. But we can understand the universe. That makes us something very special.[111]

---

110. Piet Hein, *Grooks*, MIT Press.
111. Dennis Overbye, *Stephen Hawking Dies at 76; His Mind Roamed the Cosmos*, March 4, 2018.

Summary

Homo sapiens, an intelligent, sentient hominin, has evolved the mental capabilities to study Nature. In this pursuit he (she) has made scientific discoveries that give insights into Nature's creation of the universe and biological life on Earth. Although currently modest in extent, the available scientific information is adequate to outline an unfolding path of creation from the beginning of the universe—from the big bang to *Homo sapiens* and his inventions of gods. He has evolved not only the capability to think scientifically about his creation but also the capability to invent supernatural stories about creation with gods.

As would be expected, these two creation descriptions differ; the scientific view holds that Nature operates under its set of laws and is the cause of creation, while the stories conceived by man hold that the supernatural gods who operate with no reference to Nature's laws are the creators of the universe and man.

The authority of the scientific community is based on Nature's laws which have been discovered through observations and experiments conducted by scientists. These have led to the formulation of mathematical theories that describe Nature. Mathematics is the language of Nature and is used to communicate the knowledge of Nature's creations.

The authority of the Christian God is based on the Christian narrative derived from religious texts. However, there are many religions with other gods, some of which conflict with the Christian one. Which god is *the* god is not argued. The Christian God is the reference God used.

Two central stories in the Christian narrative describe the supernatural creation of the universe and of the first man and woman, Adam and Eve, plus a talking serpent. These stories are unsupported by facts and conflict with Nature's version of the evolutionary history of the universe, Earth, and the biological tree of life on Earth. In Nature's view, *Homo sapiens* (Adam and Eve, so to speak) evolved from a long line of vertebrate species over 500 million years.

Gods made their appearance in man's evolutionary journey after *Homo sapiens* evolved the mental capability to conceive supernatural spirits and gods about 50,000 years ago. Since then, many religions—polytheistic and monotheistic—have been conceived by *Homo sapiens* with narratives of gods and creation stories that conflict with Nature's laws and the laws of the many other religions.

The authority of religions and their god narratives conceived by *Homo sapiens* are not based on the language of mathematics, which was not available to humans thousands of years ago when religions and their gods were conceived. Accordingly, and unsurprisingly, their narratives describing creation are not supported by Nature's mathematical laws. Without a mathematical set of laws, an unbridgeable gap exists between Nature's description of the creation of the universe and the many religious supernatural creation stories.

However, religions—and there are many—have played an important social role in man's evolution in that they have served and continue to serve as sources of social concepts that are helpful in fostering group togetherness, personal well-being, and survival of *Homo sapiens.*

One of the early religions is Judaism, a religion that solidified its monotheistic God concept hundreds of years before the Common Era. From Judaism, Christianity emerged and defined its concept of God hundreds of years later in the early Common Era. Like all religions, they have many internal denominations that continue to redefine their God.

The first Christian religious narratives and laws were written over 2,000 years ago and, once written, they were declared to have been given by God and were deemed holy and not subject to change. Beliefs in religions do not require acceptance of Nature's laws, so when new scientific data appear, the old religious narratives and laws are not corrected and become out of date and, accordingly, pose problems for believers when they encounter new facts of the natural world.

It has been most difficult for religions to avoid such conflicts with emerging scientific knowledge, for religions have had great social power that has enabled them to inject their beliefs outside their discipline over the last 2,000 years, while the scientific community has had little power. This imbalance of power has led to an overreach by religious authorities to impose their laws on Nature's laws as well as on state laws and cause many conflicts.

The concepts of gods, with their associated godly and social laws, have been and will continue to be important to the well-being of believers. Supernatural god concepts have given emotional support to believers in the face of life's mysteries, stresses, and uncertainties. State laws have given social frameworks for groups and states to function without conflict; however, understanding Nature's laws is necessary for man's advancement and survival. Living under these three sets of laws—those of Nature, the state, and religion—requires a clear separation of Nature's laws from those of the state and religion to minimize conflicts with religious authorities elevating their God over the authorities of Nature and the state.

A hierarchy of information is needed to reduce conflicts:

Summary

Nature's laws are honored over those of the state and gods,
State laws are honored over god's laws, and
God's laws are for believers.

There are many examples of conflicts when these rules are not followed. One is the belief that god is involved in human births. By placing god's laws over Nature's laws religious tribes wishing to grow are unimpeded in population growth. Scientific knowledge of births is not being used. As the world's population continues to grow, the negative impact on the biosphere of the growing numbers of people will also continue to grow.

The social skills that have brought *Homo sapiens* a multi-religion democracy in some countries seem to have stalled, unable to take the next step of growth: participating in a worldwide, multistate, multi-religion responsive body. Such an organization is required for managing pressing worldwide problems, such as increased climate change, population growth, , territorial dominance, the proliferation of nuclear weapons, and the rise of viruses and superbugs.

It is to be hoped that *Homo sapiens* will find ways for religions to join with all men and free man's extraordinary mental capabilities and social and organizational skills to address these problems and ensure the viability of the biosphere for its inhabitants in the future.

Meanwhile, scientists will continue to broaden our understanding of Nature's universe and human life on the planet Earth. Possibly, we have not even been asking the right questions about Nature or about where Nature is taking the universe. But with the little we do know about Nature; we can list probable futures of mankind within the limitations of the theories we have today (knowing that predictions of the future are always problematic). New discoveries and wrinkles in old theories will no doubt give us new insights that may change our understanding.

On humanity's scale there are many short-term problems to deal with, and we wish our descendants well. In the long term of Nature's scale, there are events that humanity cannot control, for Nature, not humanity, is in charge. One of the several projections of Nature's actions in the future is that five billion years from now, our Sun—a star—will deplete its fusion fuel and die, and in dying will cool and expand into a red giant star with a radius that swallows Earth before it collapses into a supernova explosion. Elsewhere in the universe, stars will keep forming, burning, and dying until the last stars in the last galaxies have formed, consumed their fuel, died, and had their lights extinguished. All galaxies, including our own, will become a graveyard of the cinders of planets and stars.

The trajectory on which Nature has propelled us from our big bang genesis to this point in our history we know a little about, but the path that Nature will take in the future, and that we will take with it, we know little about. We can only use Nature's laws to make estimates of future events.

In conclusion, the case for Nature's creation of an evolving, rational animal (*Homo sapiens*) who can explore and think about his universe has been outlined. But for mankind to make use of this knowledge is problematic, for most humans have social and religious values that change over time and are not found in Nature. With the high emotions and passion of humans, one cannot predict whether passion or reason will direct their decisions. For some, the passion of seeking knowledge is enough; others many want to put their passion into seeking other things— beauty, love, charity, or supernatural beliefs. When these activities are based on Nature's laws, harmony can be found. However, if denial of science provided by supernatural stories is employed, the unsupportable passions of religious laws and goals will diminish our capabilities for rational decision making. We will get older but not wiser. The choice of which path to take is up to each of us.

What we know about Nature's trajectory of our past and our possible future trajectory is briefly summarized in Figure 18, "Nature's Trajectory in a Nutshell."

Figure 18
Nature's Trajectory in a Nutshell

Nature created a big bang, the genesis of our universe, that created an expanding universe. From the dust of the bang stars were created, lived their lives, and exploded, giving more dust for stars, planets, and other cosmic entities. Around one star, a planet (Earth) formed where the dust from the stars and the Sun's warmth were used by Nature to give birth to biological life forms. These evolved into many animals, and after years of competition, one, *Homo sapiens*, evolved an intelligence with which it became the master of of the natural world with its many animals, some close cousins. Man discovered, prospered, built many things, and invented tales of supernatural gods for solace. Success led him to declare he was the tribal master. But his many tribes were poor guardians of the Earth's land, air, and oceans, which they daily poisoned because of their zeal for self-comfort, and they disregarded the production of an excess population. The Earth could be nourished by the toils of good men who enjoy their time and accomplishments or spoiled by plotting of self-serving men seeking wealth who care not for other people or Nature. Nature cares not, and moves on, for Nature was here billions of years before man and will be here billions of years after man. We have only an uncertain glimpse into our future. Our Sun may expand and engulf the Earth, the expanding universe may freeze everything, black holes may eat it all, or dark energy may rip the universe apart. Each path would be the end of life for worms, fish, birds, lions, ministers, scientists, politicians, and the rest of us with our tales of supernatural gods with stories we have invented to save ourselves. Our restless universe will continue following Nature's laws that got us here and will determine the trajectory of our future. Have a nice day.

Appendix A

Summary of Four Forces of Nature[112]

ELECTROMAGNETIC FORCE: Explains why atoms hold together and how light behaves.
Governing Theory: Quantum electrodynamics (QED)
Mediator: Photon (predicted by Albert Einstein in 1905)
Maximum Range: Infinite

WEAK NUCLEAR FORCE: Accounts for radioactive beta decay and the nuclear fusion that fuels stars.
Governing Theory: Electroweak theory (unified theory with QED at high energies)
Mediator: W and Z bosons (predicted in 1968, discovered in 1983)
Typical Range: 10–18 metres

STRONG NUCLEAR FORCE: Holds protons and neutrons together within the atomic nucleus.
Governing Theory: Quantum chromodynamics (QCD)
Mediator: Gluons (predicted in 1962, discovered in 1979)
Typical Range: 10–15 metres

GRAVITY FORCE: Keeps galaxies together, the planets moving around the sun, and our feet on the ground.
Governing Theory: General relativity
Mediator: None found; possibly gravitons posed by quantum dynamics
Range: Infinite

---

112. New Scientist, newscientist.com/article/mg24632821-000.

Appendix B

Notes on the Ten Commandments

In the course of evolution, social governance laws have been developed by tribes, societies, and states. With the arrival of religions, a new authority, god, appeared and man received a new set of laws, his religious laws, to be followed by believers. Thus, man, if he is religious, has two sets of laws to obey. Examples of these laws can be found in the Ten Commandments. They are social laws that reflect earlier social laws of Jews and later Christians.

Social laws reflect interactions or group behavior that are similar from tribe to societies, while religious laws are unique to each tribal religion. Examples of early social laws found in King Hammurabi's laws (HamLaw) and subsequently included by the Jews and later the Christians in the biblical Ten Commandments (TenCom) are noted below. The religious laws are unique to Judeo-Christian religions and, as was the case in ancient times, all were handed down by gods. The source of Hammurabi's laws is the Code of Hammurabi, which he claimed to have received from Shamash, the Babylonian god of justice. The source of the Ten Commandments laws is the Jewish Torah, which describes Moses accepting them from his God.

Social laws
Theft
- Do not steal (TenCom no. 8)
- If anyone is committing a robbery and is caught, then he shall be put to death (HamLaw no. 22)

Adultery
- Do not commit adultery (TenCom no. 7)
- If the wife of a man has been caught lying with another man, they shall bind them and throw them into the waters. If the owner of the wife would save his wife, then in turn the king could save his servant (HamLaw no. 129)

Religious laws
- Worship only God (TenCom no. 1)
- Make no graven images (TenCom no. 2)
- Do not take the Lord's name in vain (TenCom no. 3)
- Observe the Sabbath (TenCom no. 4)

The two social laws are examples of laws reflecting the humanity of the group at the time they were written over 2,000 years ago, but they remain similar to laws of today.

The four religious laws are about religious concepts specific to Judaism and its spin-off, Christianity, and they have no purchase outside of these religions.

Appendix C

Morality as Viewed by Heisenberg

The opinions about the foundations of ethics or morality of Werner Heisenberg, a renowned physicist, leader in the creation of quantum mechanics, and winner of the Nobel Prize in Physics (1932), can be taken as an example in the evolution of our understanding of morals by scientists. Late in his career, he addressed the question of religion and science, and in a speech[113] in 1974 before the Catholic Academy of Bavaria to accept the Romano Guardini Prize, Heisenberg said:

> *In the history of science, ever since the famous trial of Galileo, it has repeatedly been claimed that scientific truth cannot be reconciled with the religious interpretation of the world. Although I am now convinced that scientific truth is unassailable in its own field, I have never found it possible to dismiss the content of religious thinking as simply part of an outmoded phase in the consciousness of mankind, a part we shall have to give up from now on. Thus, in the course of my life I have repeatedly been compelled to ponder on the relationship of these two regions of thought, for I have never been able to doubt the reality of that to which they point.*

Later he wrote:

> *Where no guiding ideals are left to point the way, the scale of values disappears and with it the meaning of our deeds and sufferings, and at the end can lie only negation and*

---

113. Werner Heisenberg, *Naturwissenschaftliche und religiöse Wahrheit* [in English, "Scientific and Religious Truth"]. Taken from a speech before the Catholic Academy of Bavaria, March 24, 1973.

despair. Religion is therefore the foundation of ethics, and ethics the presupposition of life. [114]

Why would one disagree with Heisenberg's statements that religion is the foundation of ethics or morals? He was wrong, for he was simply not aware of the data on the biological evolution of morals. When he said, "Where no guiding ideals are left to point the way," he was not aware of the information that has since been gathered on the evolution of morality that indeed has given us "guiding ideals."

The evolution of morality in our ancestors and, eventually, us follows the evolution of the brain. The foundations of morals or "guiding ideals" evolved in our ancestors (chimpanzees, among others) and follow-on species before they were able to conceive of gods or religions. But with the evolution of *Homo sapiens* and their eventual conceptions of religions ideas, a second set of morals were added: godly morals. Thus, today our morals have two basic components: our evolved morals, known as secular humanistic morals, and our godly morals (commandments, laws, etc.) proposed by religious believers to serve their gods. Religious morals differ among the thousands of religions, but the secular humanistic moral component has remained somewhat constant over many religious and nonreligious societies. So, religious morals have a role for believers, as do secular humanistic morals for all of us, in establishing our "guiding ideals."

Regarding the authority of science, Heisenberg noted that "science truth is unassailable in its own field," but he did not address the question of the biblical truth of the creation of Adam and Eve as set against the unassailable science truth of the biological evolution of man. Heisenberg stayed in the field of science, but the same rules for "unassailable" truth used in science must also be applied to all information used by man, including that from religions.

---

114. Werner Heisenberg, *Physics and Philosophy: The Revolution in Modern Science,* 2007.

Index

A

accelerators
   cyclotron, 38, 39
   linear, 38
   Van de Graaff, 38
Adam and Eve, 2, 12, 83, 88, 90, 91, 92, 93, 94, 97, 107, 116
alternative facts, 6, 14, 15, 16
amendments, US Constitution, 102, 103, 104, 105
American Revolution, 11
animism, 78
Archimedes, 17, 20
Ardipithecus, 58, 60, 61
Aristotle, 17, 19, 33, 111
artificial intelligence (AI), 25
astronomy, 2, 6, 17, 21, 22, 25, 26, 32, 33, 39, 53
Australopithecus, 58, 60, 61, 62, 63, 64, 65, 67
authenticity hierarchy rule, 117
authorities for laws, 6, 7, 8, 9, 16, 117

B

Becquerel, Henri, 34
beliefs, 6, 51, 78, 84, 85, 86, 89, 90, 100, 101, 117, 119
Bethe, Hans, 36
Bible, 4, 5, 8, 12, 83, 84, 88, 92, 98, 103
big bang theory, 1, 3, 4, 5, 6, 22, 23, 29, 30, 32, 36, 40, 41, 42, 43, 44, 45, 47, 48, 90, 91, 115, 116, 119
biology, 2, 6, 22, 25, 33, 50, 53, 94, 95, 97, 107
black holes, 9, 21, 30, 31, 32, 40, 43, 44
Bohr, Niels, 34, 35
bonobos, 57, 58, 59, 60, 61, 65, 75, 88
Bruno, Giordano, 88

C

Cambrian explosion, 54, 55, 56
cave art, 72, 77
cells (biological), 7, 47, 48, 50, 54, 75, 93
chimpanzees, 2, 10, 49, 57, 58, 59, 60, 64, 65, 70, 72, 78, 88
chromosome, 50
church, separation from state, 100, 101, 102, 106, 109
Churchland, Patricia, 76, 94, 95, 99, 110
Churchland, Paul, 93
Civil War, 102, 104
Constitution, United States, 11, 16, 98, 99, 100, 101, 102, 103, 105, 106, 109
Copernicus, Nicolaus, 3, 18, 20, 21, 26, 27
cosmic microwave background radiation (CMB), 30, 42
Crick, Francis, 51, 93
Curie, Marie, 34
Curie, Pierre, 34

D

Darwin, Charles, 48, 49, 51, 69, 94
Darwin's theory, 2, 8, 18, 48, 49, 50, 51, 52, 53, 69, 88, 94, 97, 107, 108
de Waal, Frans, 59, 60, 75
Denisovans, 69, 70, 71, 73, 88
DNA, 2, 7, 25, 48, 49, 50, 51, 52, 56, 57, 58, 65, 70, 88, 97, 115
Doolittle, Russell, 97
Durkheim, Emile, 78
Dwight, Timothy, 101

E

Einstein, Albert, 2, 28, 29, 32, 34, 35, 36, 39, 44
evolution
    biological, 1, 2, 3, 4, 5, 6, 7, 8, 22, 25, 26, 43, 44, 45, 47, 48, 49, 50, 51, 52, 53, 54, 55, 56, 57, 58, 60, 61, 62, 64, 65, 66, 67, 68, 69, 71, 72, 73, 75, 76, 78, 88, 90, 91, 92, 94, 95, 97, 107, 108, 115, 116, 117
    concept of god, 3, 4, 5, 6, 8, 12, 22, 25, 26, 45, 75, 76, 78, 81, 85, 92, 116
    Earth, 1, 3, 4, 5, 6, 22, 34, 41, 43, 44, 47, 48, 53, 54, 69, 91, 118
    universe, 1, 4, 5, 22, 23, 29, 30, 32, 36, 40, 41, 42, 43, 44, 91, 92, 116, 118

F

facts, alternative, 6, 14, 15, 16
fathers, founding, 11, 100, 101
flood, biblical, 2, 12, 96
Fossey, Dian, 59
fossils
   Australopithecus
      Couple, 61, 63, 72
      Lucy, 61, 65
      Selam, 61, 64, 65
   Tiktaalik, 57
fossils, living, 56, 57, 58
Franklin, Benjamin, 28, 101
Franklin, Rosalind, 51
Friedman, Richard Elliott, 8, 83

G

Galilei, Galileo, 3, 18, 26, 33, 88
gene, 2, 48, 49, 50, 51, 70, 73, 97
Genesis (biblical), 36, 90, 91, 93
genesis (nature), 1, 3, 4, 91, 119
genome, 25, 49, 70, 73
Gilgamesh, Epic of, 12, 81, 82
God, Christian, 2, 3, 5, 6, 14, 82, 83, 84, 85, 88, 89, 91, 92, 94, 101, 103, 104, 116
God, Jewish (Yahweh), 5, 6, 12, 80, 82, 83
gods, other, 1, 3, 4, 5, 6, 8, 11, 12, 71, 75, 76, 77, 78, 79, 80, 81, 82, 84, 85, 87, 92, 95, 98, 116, 117
gorillas, 10, 57, 58, 59, 61, 66
grave goods, 69
gravitational waves, 31, 32
gravity, 9, 18, 21, 28, 31, 33, 35, 39, 40, 41, 42

H

Haidt, Jonathan, 98
Hammurabi (Babylonian king), 10, 81
Hammurabi's laws, 10, 12, 81
Hawking, Stephen, 115
Heisenberg, Werner, 19

heliocentric theory, 3, 20, 21, 26, 27
hominids, 57, 75
hominins, 23, 53, 56, 57, 58, 60, 62, 64, 65, 73, 75, 116
Homos
   Homo erectus, 66, 67, 68
   Homo floresiensis, 66, 68, 73
   Homo habilis, 62, 63, 66, 67
   Homo heidelbergensis, 66, 68, 69, 70, 71
   Homo neanderthalensis, 66, 69, 97
   Homo sapiens, 2, 3, 4, 5, 6, 8, 10, 22, 26, 43, 44, 45, 47, 48, 49, 52, 53, 56, 60, 61, 63, 64, 65, 66, 68, 69, 70, 71, 72, 73, 75, 76, 77, 78, 79, 85, 92, 95, 97, 115, 116, 117, 118, 119
Hoyle, Fred, 36
Hubble, Edwin, 28, 29

I

Inquisition, 88, 111, 112
intelligent designer theory, 88, 97, 108

J

Jefferson, Thomas, 98, 100, 101
Jesus, 82, 83, 84, 97, 98, 101, 112
Judaism, 12, 82, 117

K

Kepler, Johannes, 18, 21, 26
King, Barbara, 76
Kneeland, Abner, 88, 108

L

Large Hadron Collider, 39
Laser Interferometer Gravitational-Wave Observatories (LIGO), 31
laws
   biblical, 1, 8, 12, 13, 14, 94, 100
   Nature's, 1, 3, 4, 6, 7, 8, 9, 10, 16, 19, 21, 22, 28, 32, 38, 40, 43, 44, 50, 86, 87, 89, 94, 96, 100, 106, 107, 115, 116, 117, 118, 119
   religious, 1, 6, 8, 9, 10, 11, 12, 13, 14, 16, 24, 79, 80, 81, 83, 84, 87, 89, 95, 96, 100, 102, 103, 104, 107, 110, 113, 116, 117, 118, 119

secular, 6, 8, 9, 10, 11, 12, 14, 16, 24, 79, 80, 81, 84, 86, 87, 94, 96, 99, 100, 104, 111, 113, 117, 118
social, 11, 12, 13, 75, 78, 79, 94, 95, 96, 99, 117
Leakey, Louis, 63
Leakey, Mary, 62, 63, 64
Lemaitre, George, 28, 29, 91
life
creation (biblical), 2, 90
creation, natural evolution, 3, 4, 5, 6, 43, 44, 47, 48, 52, 53, 115, 116
language of, 25
tree of, 48, 53, 54, 56, 57, 58, 75, 88, 116
Lovelace, Ada, 21
Löwenmensch figurine, 77
Lucretius, 33, 85
Lysenko, Trofim, 106

M

Madison, James, 100, 101
Magna Carta, 99
mathematics, 1, 6, 8, 17, 18, 20, 21, 22, 26, 32, 40, 116
Maxwell, James Clerk, 7, 33
Mendel, Gregor, 51
Mesopotamia, 23, 24, 80
microscopes, 22, 25, 51
Miller–Urey experiment, 49
Mithen, Steven, 71
monotheism, 79, 80, 81, 82, 116
morality, 12, 52, 59, 60, 76, 79, 85, 90, 94, 95, 99, 104, 111
morals, 13, 53, 76, 83, 94, 95, 100, 103, 110, 113

N

natural selection theory, 2, 18, 50, 51, 52, 56, 76, 88, 97
Neanderthals, 69, 70, 71, 73, 77, 88
Newton, Isaac, 1, 2, 7, 18, 21, 27, 28, 33, 44
Noah (biblical), 12, 96

O

Obergefell v. Hodges, 104, 105
Olduvai Gorge, 62, 63

P

Paul (apostle), 83
physics, Standard Model, 39
Planck, Max, 34
polytheism, 78, 79, 80, 81, 82, 84, 116
Priestley, Joseph, 101

Q

quantum theory, 35, 39

R

relativity, general theory, 28, 29, 32, 35, 39
RNA, 25, 48, 49, 50, 51
Roe v. Wade, 104
Röntgen, William, 34
Rutherford, Ernest, 34

S

Sagan, Carl, 47
Shubin, Neil, 57
sin, 90, 93, 94, 111
social intelligence, 52, 59, 75, 76
sociobiology, 18
soul, 90, 93, 111
Standard Model of physics, 39
steady-state theory, 29, 30, 36, 43

T

technology, 23, 24, 49, 72
telescopes
   Hubble, space-based, 28, 29
   optical, 28, 29, 30, 32
Ten Commandments, 13, 81, 83, 96, 109
Thomson, J. J., 34
Torah, 5, 12, 83, 103
Torrey, Fuller, 67, 71, 72

V

van Leeuwenhoek, Anton, 51
Vatican, 3, 4, 27, 87, 88, 92, 102, 103, 105, 110, 111, 112
Venter, J. Craig, 48

W

Wade, Nicholas, 76
Watson, James, 51
waves, gravitational, 31, 32
Wilson, E. O., 18, 52, 95
Woese, Carl, 48

Y

Yahweh (Jewish God), 5, 6, 12, 80, 82, 83

www.ingramcontent.com/pod-product-compliance
Lightning Source LLC
Chambersburg PA
CBHW021953170526
45157CB00003B/972